HEADQUARTERS
DEPARTMENT OF THE ARMY

FIELD MANUAL
No. 3-21.38 (FM 57-38)

AF824809

PATHFINDER OPERATIONS

OCTOBER 2002

DISTRIBUTION RESTRICTION: Approved for public release; distribution is unlimited.

FIELD MANUAL
NO. 3-21.38

HEADQUARTERS
DEPARTMENT OF THE ARMY
WASHINGTON, DC.1 October 2002

CONTENTS

DISTRIBUTION RESTRICTION: Approved for public release; distribution is unlimited.

*This publication supersedes FM 57-38, 9 April 1993.

CHAPTER 1
INTRODUCTION

Army pathfinders mainly provide navigational aid and advisory services to military aircraft in areas designated by supported unit commanders. The pathfinders' secondary missions include providing advice and limited aid to units planning air assault or airdrop operations.

1-1. EMPLOYMENT

The pathfinders provide navigational aid and air traffic advisories for Army aircraft. This occurs at any phase of an air assault or ground operation that requires sustained support by Army aircraft. The commander employs pathfinders on a short-term basis for some missions. He can redeploy the pathfinders after they complete a major troop lift or airdrop.

a. **Primary**. Ideally, the commander assigns a pathfinder team to each combat aviation battalion. This enhances the relationship between aviators and pathfinders, who have to work well together and understand each other in order to successfully complete a mission. Aviators and pathfinders must maintain a good working relationship, despite the reduction of pathfinder units and the assignment of pathfinder-coded positions to ground units.

(1) Many units might have no trained pathfinder assets. In this case, higher headquarters must temporarily assign pathfinder assets from an external source to train supported unit personnel and oversee the conduct of pathfinder operations.

(2) Non-pathfinder-qualified soldiers receive training from the pathfinders and form into a company-level pathfinder team. Once trained, the team provides navigational aid, air traffic advisories, and any other relevant information. Around the clock, the pathfinder team supports any type of air movement or resupply operation conducted by or for the ground unit and supported by an aviation unit.

(3) Trained, equipped pathfinders select, mark, improve, and control landing sites. Engineers in direct support (DS) of lifted ground units may help pathfinders improve landing zones (LZs). In most situations, pathfinders perform two or more of these jobs at the same time. In each case, they start out by setting up ground-to-air radio communications. Also, combat lifesaver-qualified pathfinders supplement internal medical support.

b. **Secondary**. When not performing duties for supported units, pathfinders remain with their equipment, near and in communication with the supported ground unit CP. While pathfinders await further missions, the parent or supported CP may task them to help control the aviation unit base airfield, to perform minor demolition work, or, in staff sections, to perform map and aerial photographic work. However, before the pathfinders perform secondary missions, they must first train and perform routine maintenance on their equipment.

1-2. CAPABILITIES

Appropriately equipped and trained pathfinders--

a. Reconnoiter areas selected by supported unit commanders.

b. Select LZs and drop zones (DZs).

c. Infiltrate areas of operation by foot, vehicle, watercraft, or air.
d. Rappel or parachute from aircraft.
e. Prepare LZs and DZs.
f. Establish and operate visual and electronic navigation aids.
g. Remove minor obstacles.
h. Use ground-to-air (GTA) radio communications to guide pilots and advise them of air traffic within the area of operations (AO).
i. Coordinate directly with fire support units and keep pilots informed about friendly mortar and artillery fires.
j. Provide technical assistance in assembling supplies, equipment, and troops before loading the aircraft for deployment to LZs and DZs.
k. Advise and provide limited physical assistance in preparing and positioning supplies, equipment, and troops for air movement.
l. Conduct limited NBC monitoring and surveying of designated areas.
m. Provide limited weather observations, to include wind velocity and direction, cloud cover, visibility, and approximate cloud ceiling.
n. In the absence of special tactics team (STT), by agreement with the USAF, operate DZs and airfields for USAF aircraft.
o. Survey DZs for use by USAF and Army aircraft. In this situation, pathfinders might require USAF-compatible UHF or VHF radios. Aviators and pathfinders coordinate to make sure everyone knows the ground markings and radio procedures.

1-3. LIMITATIONS

When they guide aircraft or perform other, related primary tasks such as the following, pathfinders require augmentation:

- Provide security.
- Remove major obstacles.
- Recover and assemble equipment and supplies.
- Operate additional radio nets and telephones.
- Transport equipment.
- Conduct detailed NBC monitoring and surveying.

1-4. EQUIPMENT

Pathfinders use a variety of equipment. Though the aviation unit SOP may specify the type of equipment pathfinders will use, the mission dictates what specific items of equipment the pathfinders will take on the operation.

a. **Navigation Aids**. Pathfinders use navigation aids to help aviators find and identify an exact area.

(1) ***Electronic Navigation Aids***. With these aids, pathfinders can signal farther than they can with visual navigation aids:

- Homing beacons.
- Transponders.
- Radios.
- Any other electronic devices that can aid in aircraft navigation.

(2) ***Visual Navigation Aids***. With these aids, pathfinders can designate specific areas or points on LZs and DZs. They use them as GTA signals. Unfortunately, visual aids are visible, so the enemy can also see them.

(a) *Day*.
- Panels.
- Smoke.
- Signal mirrors.
- Colored gloves for signalmen.

(b) *Night*.
- Light beacons.
- Lanterns.
- Baton flashlights.
- Strobe lights.
- Pyrotechnics.

(c) *Day or Night*. Pathfinders can make field-expedient visual aids for day or night.

(3) ***Infrared Navigation Aids***. At night, pathfinders can use any infrared navigation aids that are compatible with their NVG.

b. **Communications**. Pathfinders use FM radios with secure capability and limited wire equipment. These radios allow pathfinders to communicate with aircraft, other pathfinder elements, and supported units. Incorporated homing capabilities in these radios allow pathfinders to provide navigational aid to aircraft.

c. **Assembly Aids**. Pathfinders use assembly aids to designate troop and supply assembly areas. Assembly aids include both electronic and visual devices. The pathfinders can also use or make field-expedient devices to aid in assembly. Because assembly aids can attract the enemy's attention, pathfinders must carefully avoid compromise.

(1) ***Electronic Assembly Aids***. These include radios and homing devices that work by radio signal. Enemy direction-finding equipment detects electronic signals.

(2) ***Visual Assembly Aids***. These simple-to-use aids allow positive identification of assembly areas. The enemy can also see them. To ensure understanding, pathfinders must closely coordinate the use of visual assembly aids. (See TM 9-1370-206-10, FM 21-60, and STANAGs 3117 and 3281.) Visual assembly aids include--

(a) *Day*.
- Panels.
- Smoke.
- Armbands.

(b) *Night*.
- Lanterns.
- Flashlights.
- Light beacons.
- Strobe lights.
- Chemiluminescent lights.
- Pyrotechnics.

(3) ***Infrared Assembly Aids***. Pathfinders can use infrared light sources as assembly aids, but, if they do so, both the pathfinders and the pilots must use NVDs.

d. **Miscellaneous**. Pathfinder equipment also includes--
- Vehicles.
- Binoculars.
- Night vision devices.

- Nonelectric demolition kits.
- Wind-measurement equipment.
- Parachutes.
- NBC-detection equipment.
- Thermal sights.

1-5. COMMUNICATIONS SECURITY

Pathfinders and terminal guidance personnel must know about any hostile data collection and exploitation activities. Such activities seek to disrupt, deceive, harass, or otherwise interfere with the command and control of pathfinder operations.

a. **Enemy Interception**. The enemy can intercept, analyze, determine the direction of, and exploit electromagnetic energy radiating from any signal equipment such as radios, radar, and more. He uses this intelligence for fire and maneuver and for electronic countermeasures.

(1) The enemy may collect pathfinder emissions data for immediate or later use. He may use jamming or deception, or he may continue to monitor and analyze the data for later use.

(2) Time-distance factors limit the enemy's ability to exploit signal intelligence in support of his ground operations. He may use a reaction force, or he may find the source of the signal using a direction finder. If he uses DF equipment, he may also use ECM to jam and deceive the pathfinder's electronic aids.

b. **Pathfinder Awareness**. Pathfinders must plan for the enemy's DF capabilities. Automated DF systems determine line bearings for each signal detected. The enemy continuously processes and compares line bearings and plots fixes for pathfinder signals. Depending on the size of the DF base and the number of DF systems available, the enemy may accurately locate a friendly position with little difficulty.

(1) What the enemy does to gain SIGINT reveals his intent. In combating enemy DF systems, pathfinders consider the following:

- The high priority of aviation-related missions.
- The length of time the pathfinders remain on the air.
- The number of pathfinder transmitters.
- The distance of friendly forces from--
  - Enemy DF systems.
  - Enemy fire and maneuver elements.
  - Enemy collection and jamming resources.
- Friendly actions to mask pathfinder operations.

(2) Strict signal security practices, to include EW, greatly reduce the vulnerability of signal devices to enemy exploitation. (FM 24-18 and FM 34-40 provide more information on communications.)

1-6. TRAINING

Personnel qualify as pathfinders only by completing the pathfinder Course at the US Army Infantry School, Fort Benning, Georgia.

a. **Goals**. The pathfinder training program stresses--

(1) The development of individual proficiency in air traffic control procedures.

(2) That pathfinders learn and know the SOP of the aviation unit they support.

THIS CHAPTER IMPLEMENTS STANAGs 2863, 3117, AND 3570

CHAPTER 2
OPERATIONS

Pathfinders conduct many different missions. Several of these supplement the ground unit's operation.

Section I. PLANNING

To ensure success of the ground mission, pathfinders plan their own missions in detail. The more time they have to plan, the more detailed a plan they can make.

2-1. WARNING ORDER

As soon as he receives word of a pending operation, the senior pathfinder issues a mission alert. He immediately follows with a warning order. He includes just enough information to allow the other pathfinders to start preparing for the operation. This includes—

- Roll call.
- Enemy and friendly situations (in brief).
- Mission.
- Chain of command and task organization.
- Individual uniform and equipment (if not discussed in the SOP).
- Required equipment.
- Work priorities (who does what, when, and where).
- Specific instructions.
- Attached personnel.
- Coordination times.

2-2. INITIAL PREPARATION

On receiving the alert or warning order, pathfinders inspect and, as needed, augment personnel and equipment.

a. Pathfinders prepare equipment in the following order, from the most to the least important:

(1) Radios.
(2) Navigation aids (electronic and visual).
(3) Weapons.
(4) Essential individual equipment.
(5) Assembly aids.
(6) Other items, as needed.

b. The pathfinder element leader (or his representative) and the air mission commander begin coordinating with the supported aviation unit(s), ground unit(s), or both.

c. As the pathfinders receive more information, they reorganize personnel and equipment to better accomplish the mission. If time permits, they rehearse. They use available briefing aids, and they rehearse on terrain that most nearly resembles the AO.

d. To succeed, an operation must have security. So, each person receives only the information he must have to complete each phase of the operation. For example, the commander isolates any soldiers who know the details of the operation. The situation dictates the extent of security requirements.

2-3. COORDINATION

Ground and aviation commanders work together to coordinate and plan the details of operations for which they require pathfinder assistance.

a. In any type of operation (combat assault, reinforcement, artillery displacement, resupply, or evacuation), the pathfinders might have to recommend—

- Exact locations for DZs or LZs.
- A time schedule.
- Landing formations.
- Employment techniques.

b. Before selecting a DZ or LZ, the supported unit commander considers METT-T factors. He also considers what the pathfinder and aviation commanders (or their representatives) suggest.

c. While preparing for an operation, air liaison officers (ALOs) and ground unit commanders (GUCs) coordinate with pathfinders to make an air movement table. Table 2-1 shows who coordinates what.

SEQUENCE	ALO	GUC	COORDINATE WITH PATHFINDERS
1.	X	X	Operational location (coordinates).
2.	X		Locations of the primary and alternate communications checkpoints (coordinates).
3.	X		Location of release point.
			• Coordinates.
			• Whether manned or unmanned.
4.	X	X	Time the site can begin operating.
5.	X		Aircraft information.
			• Formation.
			• Time interval.
			• Time of flight.
			• Drop speed.
			• Drop altitude.
6.	X		Pathfinder transportation and time available for briefing.
7.	X		Pathfinder transportation station time.
8.	X	X	Routes into the objective area.

Table 2-1. ALO and GUC coordination of air movement table.

SEQUENCE	ALO	GUC	COORDINATE WITH PATHFINDERS
9.	X	X	Call signs.
			• Aircraft.
			• Pathfinders.
			• Supported units.
			• Other friendly units.
10.	X	X	Primary and alternate frequencies.
			• Aircraft.
			• Pathfinders.
			• Supported units.
			• Other friendly units.
			• Homing beacon.
11.	X	X	Fire support.
			• Artillery.
			• Tactical air support.
12.	X		Weather forecast.
			• Ceiling.
			• Visibility.
			• Temperatures (high and low).
13.	X	X	Logistical support, including locations of-
			• Medical aid station.
			• Prisoner collection point.
			• Fuel.
			• Ammunition.
			• Rations.
14.	X	X	Alternate plans (ALO and GUC).
			• Evacuation plan.
			• Escape and evasion.
15.	X	X	Friendly unit locations.
16.	X	X	Authority to implement mission change.
17.		X	Support personnel required.
18.	X		No-land or no-drop signals (day and night).
19.			Markings for obstacles (only on request of flight commander).
20.	X		Marking of objective site for identification from the air.
21.	X	X	Time allowed for approval.

Table 2-1. ALO and GUC coordination of air movement table (continued).

d. The pathfinder needs this information because he helps coordinate planning. He uses the information to make final plans for the pathfinder phase of the operation. To make sure that he can safely and efficiently control all aircraft in and around the DZs or LZs, he must know all about the operation's air movement phase. Aviation and ground commanders inform pathfinders of all changes to plans and landing sites, and about any emergencies. The pathfinder coordinates all activities with every agency or unit involved, then gives the information to all of the pathfinders involved in the operation.

e. When the pathfinder reaches the objective site, he may find it unsuitable. He evaluates the coordinated landing formation, heading, drop altitude, and the ground site itself. Then, he coordinates with the ground commander, aviation commander(s), or both to see whether any of the original requirements have changed. Depending on the METT-T situation, the commander(s) determines what, if any changes to make in order to accomplish the mission. If for any reason he cannot contact the GUC or aviation commander, the pathfinder can also coordinate for authority to change requirements.

f. The pathfinder limits augmentation to that appropriate to the amount and type of transport. The reinforced pathfinder team remains under the command of the pathfinder leader, who is responsible for team functions. Based on the coordinated plans for the operation, the pathfinder requests augmentation in personnel and equipment. He considers—

(1) Mission.

(2) Use of personnel and equipment for security.

(3) Requirement to help assemble the supported units' personnel, supplies, and equipment.

(4) Need for assistance in removing obstacles.

(5) NBC survey or monitoring requirements.

(6) Assistance required to transport and operate navigation aids under pathfinder direction.

2-4. LINKUP WITH SUPPORTED UNIT

Pathfinders join the supported unit early enough to allow final coordination between pathfinder, aviation, and lifted ground unit representatives. Pathfinders designated to accompany and provide continuous support to a ground unit can enter a DZ or LZ ahead of the assault echelon. After the initial phase of the air movement, they link up with the supported unit.

2-5. FINAL PREPARATIONS

The pathfinder leader issues his OPORD. If he issues it before linking up with the supported unit, he issues any changes as a FRAGO. The order describes any member's duties not covered in the unit SOP. Team members must have a chance to study maps, aerial photos, and terrain models of the area. The order provides details about the location and operation of proposed air delivery facilities, flight routes, flight formations, time schedules, RPs, and CCPs.

a. The pathfinder conducts a final, thorough check of equipment. The commander decides exactly how to transport the equipment into the objective area. Then the pathfinders prepare all of the equipment for rapid displacement.

b. Just before departure, at a final weather and operational briefing, the pathfinders and supported units conduct final coordination.

Section II. ORGANIZATION FOR COMBAT

The pathfinder mission itself determines the specific requirements of the mission. In most operations, three to six soldiers comprise the average-sized pathfinder element supporting a DZ or an LZ or continuously supporting an infantry battalion. Seldom does a pathfinder section deploy as a unit from a single location. The pathfinder leader plans for his elements to operate widely separated and disconnected.

2-6. INSERTION

Pathfinders can insert by a variety of air, sea, or land transportation modes.

a. **Helicopter Delivery**. Helicopters can deliver more personnel and equipment in a better state of operational readiness than any other means. Even in marginal weather, helicopters allow more precise, flexible deliveries than parachutes do. Some terrain does not allow helicopter landings. In these cases, trained soldiers rappel from helicopters, while the helicopters hover over the unsuitable landing areas. Personnel can insert or withdraw by ladders suspended from hovering helicopters. Helicopters can also—

- Furnish a means of aerial radiological monitoring.
- Rapidly shift or evacuate pathfinders.
- Carry nonparachutists to support pathfinders.
- Deliver when rain or low ceilings prohibit parachuting.

b. **Parachute Delivery**. Parachute delivery by fixed-wing aircraft normally affords greater range and speed of movement than landing by helicopter. In a short-distance operation, helicopters can serve as the jump aircraft.

(1) Depending on wind conditions, pathfinders should compute their desired parachute RPs before arriving over the DZ. For accuracy and security, the pathfinders jump at the lowest practical altitude. Aircraft SOPs prescribe jump altitudes and personnel procedures. Such procedures vary IAW peacetime and wartime restrictions.

(2) As highly trained parachutists, pathfinders can insert into unimproved and marginal DZs. They know how to control the canopy of a maneuverable parachute, and they know how to make emergency landings. They also know how to parachute into rough-terrain DZs. These skills give them some flexibility in planning parachute delivery.

> **CAUTION**
>
> During preparation for the operation, pathfinders carefully arrange and pad all essential items of operational equipment into appropriate containers. Carrying this equipment with them when they insert ensures they have it as soon as they land.

(3) The best time to insert by parachute is during nonilluminated, nonsupported night operations when the operation emphasizes secrecy.

(4) Because fixed-wing aircraft need large, secure, obstacle-free landing areas, they seldom deliver pathfinders.

c. **Water and Land Delivery**. Delivering pathfinders by watercraft offers security only up to the point of debarkation from the craft. The pathfinders still must move from the landing point (debarkation) to their final destination. To do this, they infiltrate by land.

2-7. OVERLAND MOVEMENT

Because it limits small elements to short movements, infiltrating by land is the worst way to insert pathfinders.

a. A well-organized, stable, close-knit enemy defense in depth can prohibit land infiltration. When time allows, the pathfinders can increase infiltration security by combining overland infiltration with parachute or airlanded infiltration.

b. Pathfinders infiltrate overland when the following conditions exist:

- Limited visibility over difficult terrain.
- Overextended enemy lines.
- Fluid combat zone.
- Unsecured portions of enemy boundaries.

2-8. STAY-BEHIND OPERATION

In a stay-behind operation, pathfinder elements remain in the operational area while another friendly force withdraws from the area. The commander can use stay-behind operations for the following reasons:

a. To lure enemy forces into a vulnerable position.

b. To hold an area for reoccupation. If the commander plans to reoccupy a friendly area he knows the enemy could overrun, he leaves a stay-behind force to hold it. If he leaves the stay-behind force for this reason, then he must also plan an air assault to retake the area, if needed.

Section III. CONDUCT OF OPERATIONS

Pathfinders provide air traffic advisories and navigational aid for airplanes and helicopters. They also perform limited physical improvement and NBC monitoring and surveying within DZs or LZs. Pathfinder availability, the tactical plan, the complexity of the operation, the terrain, and the air assault proficiency of the supported ground unit dictates pathfinder support. However, every air assault operation requires positive aircraft control. During an air-assault operation, pathfinders cross load before entering an LZ with the initial assault elements.

2-9. DAYLIGHT ASSAULT

In daylight operations, pathfinders insert into an LZ before the initial assault echelon only if the LZ requires extensive improvement, or if planners expect unusual control problems. Either way, the pathfinders start setting up at once so they can provide air traffic control and other aid to all subsequent lifts of troops, supplies, and equipment. They may have a few minutes or several hours to do this before the other elements arrive. The tactical plan spells out exactly when and how the pathfinders will enter the area, whether they will go in alone or not, and what time the next element will arrive.

2-10. NIGHT ASSAULT

Security and operational requirements determine the method of delivering pathfinders at night. Pathfinders can move cross-country on foot, airdrop onto or near objective areas, airland in total blackout, or airland with minimum natural illumination. When they do insert this way, they sometimes arrive before the main body does. As soon as they arrive, the pathfinders reconnoiter the LZ, install visual and electronic aids, and establish air traffic control. Soldiers from the supported ground unit sometimes accompany the pathfinders. These extra soldiers provide security and help clear obstacles. The on-site pathfinder element remains concealed and observes the objective. Pathfinders analyze the planned landing formation, heading, and assembly area. To avoid compromising the mission, no one on the DZ or LZ moves until an incoming aircraft reaches the CCP.

2-11. EXTRACTION

As the ground force at the LZ shrinks, vulnerability to attack increases. Therefore, the commander has pathfinders speed up the air assault extraction operations.

a. Planned artillery fires and air strikes as well as the need to maintain ground security to the last minute require that ground controllers control supporting aircraft throughout the extraction. This means they make sure aircraft land at specific points within the extraction site where ground security can cover them. This speeds the operation and helps ensure the safe withdrawal of personnel, equipment, and aircraft from the area.

b. Unless they land with the lifted unit, pathfinders must arrive at the extraction site in time to reconnoiter thoroughly and coordinate with the lifted unit.

c. During the planning stage, the pathfinder team leader designates near and far rally points for use in case the DZ or LZ becomes unusable. Pathfinders may have to fight their way to these rally points and reorganize. To increase the chance of survival, evasion, resistance, and escape, the team leader designates far rally points several kilometers from the DZ or LZ.

2-12. STAGING AREAS

In staging areas, in the absence of ATC units, pathfinders can provide air traffic advisories. They may also act as liaison between the aviation and ground units and help the ground unit commander prepare and position troops, supplies, and equipment for air movement. When pathfinders must set up a temporary staging area to support an operation of short duration, they move into the area before the operation begins. This gives them enough time to reconnoiter, mark the site, coordinate, and set up positive ATC. Safe, efficient, and rapid movement of helicopters or airplanes requires positive ATC in staging areas. The need for positive ATC increases when the weather deteriorates, when the number of aircraft increases, or when changes in the situation or plans require it.

2-13. ARTILLERY DISPLACEMENT

Pathfinders should help safely and rapidly displace artillery, day or night. Coordinating with ground and aviation unit commanders and understanding their SOPs ensure pathfinders accurately and efficiently deliver equipment, personnel, and ammunition.

2-14. SUPPORT OF GROUND OPERATIONS

During ground operations that require sustained Army aviation support, pathfinders might continuously aid and control aircraft. The commander can attach pathfinders he has already attached to infantry battalions to companies as well. The pathfinders provide support consistent with the availability of personnel and equipment. Continuous support improves operational efficiency and aviation safety during all types of air assault operations. However, aviation units with limited pathfinder resources cannot provide continuous support. In such cases, commanders usually employ pathfinders on a short-term, priority basis wherever the pathfinders can help accomplish major unit missions. In the absence of pathfinders, selected personnel in the ground units must receive enough training and preparation to allow them to provide minimum aid to supporting aircraft.

2-15. SUPPORT OF AIR FORCE

By joint US Army and USAF agreement, in the absence of USAF STTs, Army pathfinders can provide day or night control for USAF aircraft on airfields, DZs, and LZs. However, the USAF might need to provide pathfinders with UHF and VHF communications equipment compatible with USAF aircraft.

2-16. MIXED OPERATIONS

Some situations could require the simultaneous control of mixed air traffic at the same location such as resupply parachute drops into forward helicopter LZs. Fixed-wing airfields can expect helicopter traffic. Mixed air traffic often presents difficult control problems, so controllers must apply strict control measures. To ensure control, they designate, coordinate, and clearly identify landing, parking, loading, unloading, refueling, and rearming areas.

2-17. RADIO COMMUNICATIONS

For success, a pathfinder requires the essential element of communication by GTA voice radio. The pathfinders place this into operation first at a DZ or LZ, and they take it out of operation last.

a. Pathfinders must thoroughly understand radio procedures. This includes the phraseology unique to ATC (Chapter 3). They must send clear, concise, applicable, accurate, and correctly timed communications. To achieve speed and clarity of transmission, pathfinders and aviators practice radio discipline. They transmit only necessary messages. Also, except in emergencies, they use pathfinder ATC frequencies only for ATC (Figure 2-1).

b. Because they exchange a lot of vital information, aircraft crews normally record the important parts of GTA messages. This helps them to make sure they understand and can follow instructions.

c. Pathfinders use electronic homing beacons, visual aids, and arm-and-hand signals to complement voice communications. Pilots and transported troops must know the purpose and meaning of the aids used and the techniques for using them (STANAG 3570). (FM 21-60 discusses arm-and-hand signals and visual aids.)

d. When possible, to keep informed about changing situations that could influence their operations, pathfinders monitor supported unit command radio nets.

e. Pathfinders set up positive communications between pathfinder ATC facilities and collocated fire support elements. This ensures aircraft receive timely and accurate information about friendly fires.

f. Pathfinder operations require the constant use of radios. This gives the enemy force many chances to intercept, analyze, and exploit friendly transmissions. They try to gain intelligence and conduct electronic jamming and deception. Defeating enemy jamming or imitative deception methods falls mostly to the radio operator. He must know how to recognize and report this deliberate interference. To plan and execute a tactical mission, he must know—

- How to defend against and beat ECM using ECCM.
- How to secure transmissions.
- How to communicate using other means.

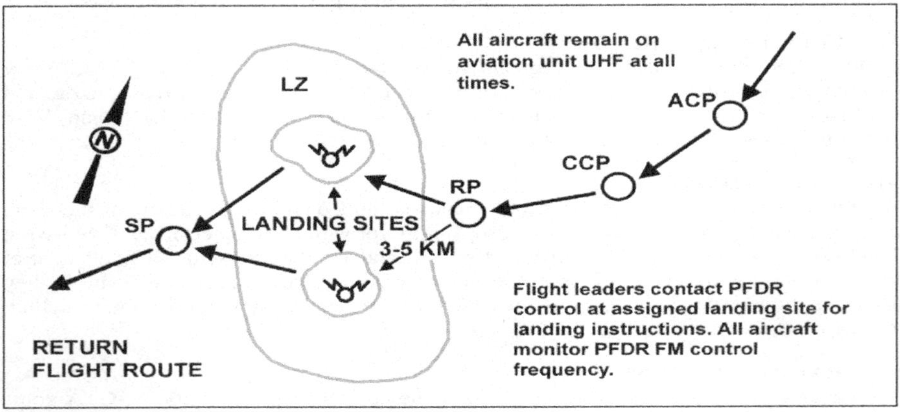

Figure 2-1. En route communication procedures with pathfinders in an LZ.

2-18. TERMINAL GUIDANCE BY SUPPORTED UNITS

Terminal guidance refers to information and minimal guidance given to pilots by anyone in a ground unit other than a qualified pathfinder. Selected personnel normally furnish terminal guidance within the supported unit. To do so, they use both organic and improvised equipment.

a. When pathfinders accompany ground units, aviation unit SOP may direct that terminal guidance personnel augment pathfinder elements.

b. Terminal guidance personnel should know—

(1) The supporting aviation unit SOP.

(2) How to operate electronic and visual navigation aids to help aircraft find DZs or LZs.

(3) To provide essential information through GTA radio to guide and control Army aircraft.

(4) To reconnoiter and recommend suitable DZs or LZs.

(5) To determine, recommend, or perform ground-clearing pioneer work to prepare DZs or LZs.

Section IV. HIGH-THREAT ENVIRONMENT

The threat comes in many forms. To ensure mission success, aviation and ground commanders must consider all possibilities. They must consider a threat anything that could disrupt or delay the mission, or that could otherwise cause the mission to fail.

2-19. CONTROL AND NAVIGATION

Pathfinders have limited voice control of aircraft. Thus, the ground unit commander and the air mission commander must coordinate closely. Navigation presents special problems-aviators must fly low to avoid detection. In a high-threat environment, critical factors include time, distance, routes, and tactical instruments.

a. For example, two pathfinders with beacons could emplace along a route in advance of the initial flight. They provide pilots with air control points. If the pilot needs the beacon turned on due to navigational error, he can transmit a prearranged signal or code word.

b. Pathfinders might discover a threat, such as an antiaircraft weapon, along the primary route. If so, they can alert pilots by prearranged code word or signal to change to an alternate route.

2-20. TACTICAL INSTRUMENT FLIGHTS

Flying under instrument meteorological conditions poses special problems in a high-threat environment. This threat overrides the controlled instrument flight rules in the aviation-series manuals. It forces aircraft to fly at altitudes well below the minimums for normal instrument flight.

a. Weather variances can create a tactical emergency. If so, the commander might have to use aviation assets under instrument conditions and well below the altitudes specified by standard instrument flight rules. The commander will only send aircraft on a mission in a high-threat environment under these conditions when the situation meets the following criteria:

(1) The aviation and ground commanders cannot postpone the mission in order to wait for better weather.

(2) The pathfinders must conduct the mission in a high-threat environment.

(3) Low visibility en route precludes nap-of-the-earth flight.

b. Aviation and ground commanders employ tactical instrument flight whenever weather or time and distance considerations prevent mission completion in other flight modes. Therefore, they must often use tactical instrument flight during round-the-clock operations on the high-threat battlefield. Air crews and pathfinders must rehearse tactical instrument flight until they achieve proficiency.

c. Pilots fly in one of two altitude modes:

(1) **Mode 1**. When the air defense threat keeps flight altitudes below those established by AR 95-1 (for standard instrument flight), then pilots can fly at least 1,000 feet over mountainous terrain and 500 feet over flat terrain.

(2) **Mode 2**. When the threat limits flight altitudes to the least possible clearances, pilots can fly as low as 50 to 500 feet above the ground, regardless of terrain.

2-21. AIR ROUTES

Aircraft traffic management personnel (and pathfinders) can expect to move their equipment as often as every four hours, depending on the threat. Terrain, weather, and, most importantly, whether or not the enemy could intercept friendly aircraft from that location, determine when to move.

a. **Threat and Terrain**. In many instances, the threat and terrain prohibit a straight-line flight between the takeoff (liftoff) point and the destination (Figure 2-2). This applies to both Modes 1 and 2.

b. **Flight Monitoring and LZ Approach**. Enemy presence keeps the pathfinders from using nondirectional beacons. However, for pilots to approach and land on the LZ visually, they need good visibility. Using radio homing signals for directional guidance presents a dubious option. Whether or not aircraft traffic management personnel decide to use this electronic device, they should try to orient its signal away from the FEBA. This reduces the chance of detection.

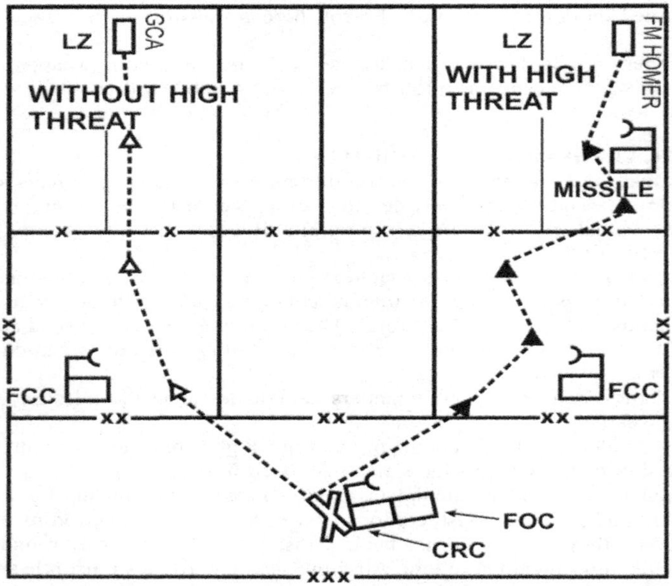

Figure 2-2. Comparison of air routes with and without a high-threat environment.

(3) Mission accomplishment in an EW environment.

b. **Commander's Responsibilities**. Major unit commanders who use pathfinders bear the responsibility for sustaining the pathfinders' training and proficiency. Pathfinder training works best when integrated with the training of supported aviation and ground units.

c. **Pathfinder's Responsibilities**. The assigned, qualified, and trained pathfinder must ensure that any nonqualified soldiers assigned to his unit team receive adequate pathfinder training before going on a mission.

CHAPTER 3
AIR TRAFFIC CONTROL

This chapter discusses the pathfinder air traffic controller, not the regular air traffic controller. The latter has radar and other sophisticated tools to monitor weather and guide aircraft. The pathfinder has only his training, so he can only advise and inform the pilot. Based on what the pathfinder tells him and on his own observations, the pilot must then decide whether to land, take off, or drop equipment or personnel.

Unless clearly stated otherwise, all mention here of "air traffic controller" refers to the pathfinder air traffic controller. Also, this chapter includes terms that are peculiar to ATC tasks.

Section I. PATHFINDER AIR TRAFFIC

The pathfinder air traffic controller uses radio or directional light signals to provide flight information, expedite traffic, and prevent collisions.

3-1. SAFETY

Pathfinders issue specific commands regulating vehicles, equipment, or personnel in the movement area. They help with search-and-rescue operations (STANAGs 2863 and 3281). They also promote the safe, efficient flow of air traffic by issuing clearances instructions, and information.

a. Pathfinders survey all visible air traffic operating within and around the airspace of the LZ, DZ, or airfield. They also bear the responsibility for all aircraft, vehicles, and personnel in the movement area of the LZ, DZ, or airfield.

b. Pathfinders acting as air traffic controllers (ATCs), provide control service by observing or knowing of traffic and airfield conditions that might constitute a hazard. These include--

- Surface conditions.
- Parachutists within control zones.
- Vehicular traffic.
- Temporary obstructions on or near the LZ, DZ, or airfield.
- Other aircraft.
- Enemy or friendly activities.

3-2. VOICE CONTROL

To communicate vocally, pathfinders and pilots must speak clearly and listen to each other. A clear, decisive tone of voice indicates control of the situation. Pilots may not trust instructions delivered in a vague or hesitant voice. To ensure that traffic flows safely and smoothly, the pathfinder must speak firmly and confidently, using standard words and phrases. Pathfinders use the phonetic alphabet to indicate single letters or initials, or to spell words, whenever similar sounds or difficulties in communication require them to do so.

a. Voice transmission offers a brief, concise, uniform flow of communication. The pathfinder controller must speak distinctly and pay special attention to numbers. When the accuracy of a message is doubtful, he repeats the complete message or essential parts.

Radiotelephone communicators use the following techniques to ensure clear understanding:

(1) Speak directly into the microphone.

(2) Speak in a normal, conversational tone.

(3) Vary your pitch-avoid speaking in a monotone.

(4) Speak at a comfortable speed-avoid speaking too slow or too fast.

(5) Keep your tone clear, professional, and firm. Avoid showing fear, indecision, anger, or other negative emotions in your tone of voice.

(6) Speak with confidence, especially in emergencies.

b. The pathfinder ATC must transmit messages only as necessary for control or help ensure safety. Specific procedures and control techniques vary, but the following rules apply regardless of the techniques used:

(1) The pathfinder issues instructions and information about all known traffic conditions.

(2) The pilot uses at lease one component of a standard traffic pattern (final approach), consistent with the pathfinder's instructions.

(3) The pilot has the final authority about whether or not to accept clearances issued by a controller.

3-3. FORMATS

A pathfinder controller uses the following formats and sequences for ground-to-air radio communication.

a. The pathfinder controller initially calls up an aircraft as follows:

(1) Identifies the aircraft he wishes to call.

(2) Says, "THIS IS."

(3) Identifies the calling unit.

(4) Identifies the type of message to follow (when this will help the pilot).

(5) Says, "OVER."

Example: TANGO TWO SIERRA TWO SIX (T2S26), THIS IS CHARLIE THREE DELTA THREE SIX (C3D36) (short pause), OVER.

b. The controller replies to an aircraft's initial call-up in this sequence:

(1) Identifies the aircraft initiating the call-up.

(2) Says, "THIS IS."

(3) Identifies the pathfinder control unit.

(4) Says, "OVER." After establishing communications with an aircraft, shortens the transmission by using only the last three numbers (or letters) of each party's (his and the aircraft's) identification.

Example: SIERRA TWO SIX, THIS IS DELTA THREE SIX, OVER.

c. The controller always starts a clearance (instruction) intended for a specific aircraft by identifying that aircraft. If he thinks that using the shortened identification could cause or is causing confusion, he can go back to using the full identification.

Example: SIERRA TWO SIX, WIND CALM, CLEAR TO LAND, OVER.

d. The controller can omit "THIS IS" from the reply.

Example: SIERRA TWO SIX, DELTA THREE SIX, OVER.

e. The controller can omit the facility identification.

Example: SIERRA TWO SIX, TURN TO HEADING ZERO FOUR FIVE, OVER.

f. Right after call-up, without waiting for the aircraft to reply, the controller can send a short message that he expects the aircraft to receive.

Example: SIERRA TWO SIX, EXTEND DOWNWIND, OVER.

g. If the message obviously requires a reply, he can omit "OVER."

Example: SIERRA TWO SIX, WHAT IS YOUR LOCATION?

h. To distinguish between similar aircraft identifications, he may emphasize appropriate numbers, letters, or words.

i. The controller never transmits to an aircraft during the final approach, touchdown, landing roll (touchdown), takeoff (liftoff), initial climb, or turnaway from the field. At these times, the pilot must concentrate on flying the aircraft. However, he transmits at once any condition or information that could affect the safety of the aircraft. Under no circumstances does the controller withhold from the pilot of an approaching aircraft any information about hazardous runways, fields, weather, or traffic conditions.

3-4. NUMBERS

A pathfinder controller transmits numbers by units or digits (Table 3-1, page 3-4).

3-5. PHRASES AND TERMS

A pathfinder controller uses particular phrases (Table 3-2, page 3-5) and terms (Table 3-3, page 3-6) to control and communicate with aircraft. He must know these phrases and how to use them.

Section II. LANDINGS

The safe landing of aircraft requires control of the airspace and grounds around the site. Managing air traffic involves using traffic patterns and maintaining separation of aircraft.

3-6. TRAFFIC PATTERNS

The pathfinder uses a traffic pattern to help manage airspace over his location, that is, in and around a landing site, airfield, LZ, or DZ (Figure 3-1, page 3-7). A traffic pattern normally extends out 1 mile from the final approach of the landing area in all directions, depending on the type of aircraft or size of the facility.

TO TRANSMIT	SAY
CEILING HEIGHTS and FLIGHT ALTITUDES	"CEILING FIVE HUNDRED" (one unit) or "CEILING FIVE ZERO ZERO" (digits for emphasis). "ALTITUDE ONE THOUSAND THREE HUNDRED" (two units) or "ALTITUDE ONE THREE ZERO ZERO" (in digits).
TIME	Use the word TIME followed by the number. For example, "0115 HOURS TIME, ZERO ONE ONE FIVE" or "1315 HOURS TIME, ONE THREE ONE FIVE."
ELEVATION NUMBERS	Use the words FIELD ELEVATION and the number. For example, for a 17-foot elevation, say, " FIELD ELEVATION SEVENTEEN." For a 50-foot elevation, say, "FIELD ELEVATION FIFTY."
WIND SPEED	Use the word WIND followed by compass direction and velocity (knots). For example, "WIND TWO SEVEN ZERO AT FIVE."
HEADING	Use the word HEADING followed by compass numbers (degrees); omit the word DEGREES. For example, "HEADING ONE TWO ZERO," "HEADING ZERO ZERO FIVE, " or, "HEADING THREE SIX ZERO." (The latter indicates a North [direction] heading.)

Table 3-1. Numbers transmitted by units or digits.

a. In a normal (left) traffic pattern, the aircraft makes only left turns. The pilot keeps the airfield, landing site, LZ, or DZ to his left. In a right traffic pattern, the aircraft makes all right turns. The pilot keeps everything to his right.

b. The controller uses traffic patterns to manage aircraft separation around a no-threat landing site. Rotary-wing aircraft can enter the pattern from any direction as long as they meet safety requirements. (Chapter 4 discusses fixed-wing procedures.) The height of the obstacles or aircraft requirements determines the altitude, which the controller can adjust as needed.

c. While in the traffic pattern, the aircraft flies between 1,000 and 1,200 feet (known as *civil altitude*), though this may vary depending on the nature and requirements of the mission.

3-7. METHODS OF ENTRY

An aircraft may enter the traffic pattern from any point and direction within the area around the landing strip or zone, consistent with safety requirements.

a. Fixed wing aircraft normally enter the traffic pattern in the first one-third of the closest leg, at an angle no greater than 45 degrees. Rotary wing aircraft may enter at any angle.

b. A straight-in approach might work best if it falls within safety requirements. On a straight-in approach, the aircraft must remain within 30 degrees to either side of the land heading.

INTENT	EXAMPLE
Issue takeoff, liftoff, or departure clearance when delay is undesirable.	SIERRA TWO SIX, CLEARED FOR IMMEDIATE TAKEOFF (or DEPARTURE), OVER.
Issue takeoff (liftoff) clearance when aircraft is delaying on the runway.	SIERRA TWO SIX, TAKE OFF (or DEPART) IMMEDIATELY OR TAXI OFF THE RUNWAY, OVER.
Authorize a requested straight-in approach after issuing landing instructions.	SIERRA TWO SIX, STRAIGHT-IN APPROACH (to landing strip or LZ) APPROVED, OVER.
Authorize a right-hand traffic pattern.	SIERRA TWO SIX, RIGHT TRAFFIC APPROVED, OVER.
Issue the landing sequence.	SIERRA TWO SIX, YOU ARE NUMBER THREE TO LAND; FOLLOW THREE EIGHT FIVE (aircraft identification number) ON DOWNWIND, OVER.
Instruct pilot to extend downwind leg to obtain necessary aircraft separation.	SIERRA TWO SIX, EXTEND DOWNWIND FOR TRAFFIC SPACING, OVER.
Advise pilot of information not included in landing instructions, but important to aircraft safety.	SIERRA TWO SIX, BE ADVISED WE ARE RECEIVING AUTOMATIC FIRE FROM THE EAST, OVER.
Try to establish communication with and learn the identification of an aircraft in the area.	UNIFORM HOTEL ONE, TWO MILES WEST OF BLUE STRIP, STATE CALL SIGN, OVER.
Instruct pilot to circle the LZ or landing strip.	SIERRA TWO SIX, MAINTAIN LEFT (RIGHT) CLOSED TRAFFIC, OVER.
Issue clearance to land.	SIERRA TWO SIX, CLEAR TO LAND, OVER.
Instruct a pilot on his final landing approach that his clearance to land has been cancelled.	SIERRA TWO SIX, CONDUCT GO-AROUND, OVER.
Inform pilot to continue his approach to the landing area.	SIERRA TWO SIX, CONTINUE APPROACH, OVER.
Inform pilot of observed aircraft condition upon request or when necessary.	SIERRA TWO SIX, LANDING GEAR APPEARS DOWN AND IN PLACE, OVER.
Describe vehicles, equipment, or personnel in the movement area in a way that will help pilots see or recognize them.	SIERRA TWO SIX, AIRCRAFT TO LEFT OF RUNWAY, OVER. SIERRA TWO SIX, VEHICLES ON TAXIWAY, OVER.
Describe military traffic as appropriate.	SIERRA TWO SIX, BE ADVISED HELICOPTER ON DEPARTURE END, OVER. SIERRA TWO SIX, BE ADVISED, CHARLIE HOTEL FOUR SEVEN (CH-47) ON RIGHT SIDE OF RUNWAY, OVER.
Describe the relative positions of traffic using the clock direction-and-distance method.	SIERRA TWO SIX, UNIFORM SIX, YOUR THREE O'CLOCK, FIVE HUNDRED METERS, OVER.

Table 3-2. Phrases.

TERM	MEANING
ABORT	Do not complete landing or takeoff (liftoff).
ACKNOWLEDGE	Did you receive and understand the message?
AFFIRMATIVE	Yes.
BE ADVISED	Indicates additional information is forthcoming, such as an unusual condition or hazard to flight.
BREAK	That is the end of my transmission to you. The following message is for another aircraft. OR That is the end of this part of the message. The next portion follows.
CONDUCT GO-AROUND	Do not land. Circle the landing area, and begin another approach.
CORRECTION	I gave you some incorrect information. The correct information follows.
EXECUTE	Drop personnel or equipment.
FORM YOUR OWN APPROACH.	You may enter the traffic pattern at your discretion. (Most suitable for aircraft with a sling load or for aircraft flights.)
GO AHEAD	Proceed with your message.
I SAY AGAIN	I am about to repeat my previous message.
LAST CALLING STATION	I do not know the identity of the station trying to establish communication.
MAYDAY	This is an emergency--clear the airways.
NEGATIVE	No.
NO DROP	Do not drop personnel or equipment.
OUT	That is the end of my transmission; you need not respond.
OVER	That is the end of my transmission; please respond.
READ BACK	Repeat message.
REPORT	Contact the control facility when you reach the location (or distance from the control station) that I am about to designate.
ROGER	I received and understand your transmission.
SAY AGAIN	Repeat your message.
STAND BY	Pause for a few seconds; (or) prepare to drop personnel or equipment.
STATE CALL SIGN	Identify your aircraft.
STATE INTENTIONS	Tell me your plans.
STATE LOCATION	Tell me your exact location.
UNABLE TO APPROVE	I must refuse your request.
VERIFY	Check with the originator.
WORDS TWICE	Communication is difficult; transmit each phrase twice.
WILCO	I understand and will comply.
YOU ARE UNREADABLE (BROKEN OR GARBLED)	I do not understand the transmission.

Table 3-3. Terms.

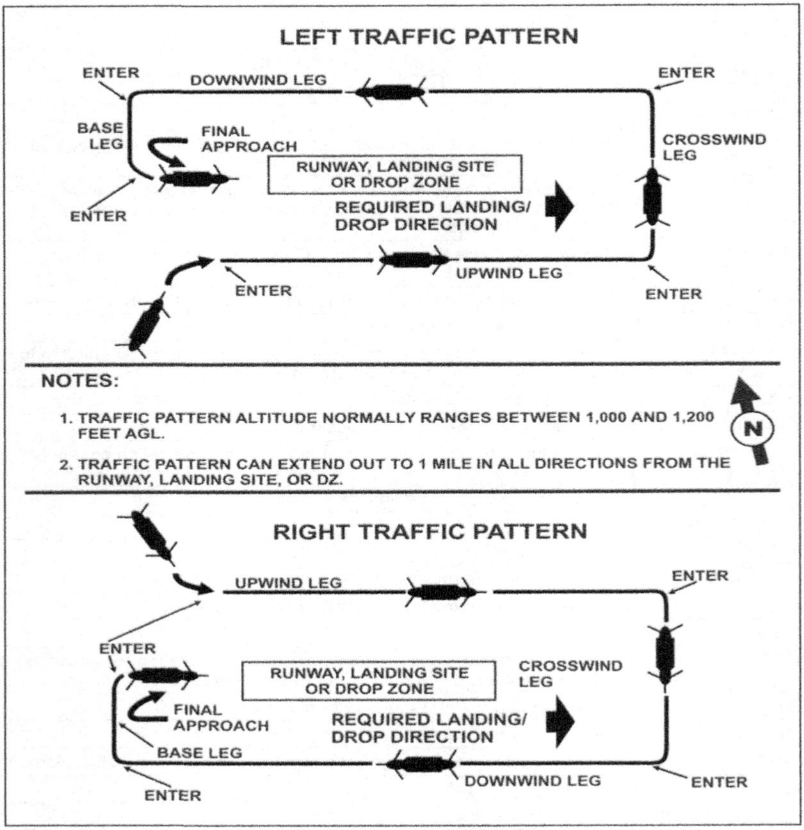

Figure 3-1. Air traffic patterns.

c. When circling to approach from any direction, the aircraft overflies the landing site then circles to the direction of landing. Normally, the pathfinder advises the pilot which direction to circle. This saves time and helps the aircraft avoid other traffic in the same airspace. The pathfinder most often encounters this type approach.

d. Departing aircraft normally leave on the same heading as landing aircraft, or as close to the same heading as they can, up to 45 degrees left or right of the land heading, depending on the wind direction. When the destination does not fall in the same direction as the departure, the aircraft may fly a portion of the traffic pattern. The pathfinder ensures that arriving and departing traffic do not conflict.

e. The controller issues a "go-around" command when for some reason the aircraft should not land after the pilot reaches the final approach leg of the traffic pattern.

f. The pathfinder uses closed traffic in either of two cases:

(1) When an aircraft does not land on the first approach.

(2) During DZ operations. When an aircraft must make more than one pass over the DZ, the pathfinder uses closed traffic.

3-8. TRAFFIC PATTERN LEGS

The traffic pattern has five possible legs. The pathfinder does not use them all at once. The pilot must at least fly the final approach leg, regardless of the type approach (Table 3-4).

a. **Upwind Leg**. This flight course runs parallel to the land heading in the direction of landing.

b. **Crosswind Leg**. This flight course runs at a right angle to the land heading, off its upwind leg.

c. **Downwind Leg**. This flight course runs parallel to the land heading, in the direction opposite of landing.

d. **Base Leg**. This flight course runs at a right angle to the landing runway off its approach, extending from the downwind leg to the intersection of the runway centerline (extended).

e. **Final Leg (Approach)**. This flight course runs in the direction of landing along the runway centerline, extending from the base leg down to the runway.

LEG	FLIGHT COURSE	DIRECTION
UPWIND	Parallel to land heading	Landing direction
CROSSWIND	Right angle to land heading	Landing direction
DOWNWIND	Parallel to land heading	Direction opposite of landing direction
BASE	Right angle to landing runway	Extends from downwind leg to intersection of runway centerline (extended)
FINAL	Along runway centerline	Landing direction; extends from base leg down to the runway

Table 3-4. Traffic pattern legs.

3-9. ADVISORY SERVICE

The pathfinder controller issues advisories for the safe operation of aircraft in his area of responsibility. He may include such information as the temporary or permanent conditions on the landing field.

a. Temporary conditions may include--

- Construction work on or immediately next to the movement area.
- Rough portions of the movement area.
- Degraded runway braking conditions due to ice, snow, mud, slush, or water.
- Parked aircraft on the movement area.

b. No two landing areas and situations are the same. Each location presents its own problems with respect to environmental conditions, peculiar weather, preferred landing directions, and so forth. For example--

(1) The final approach to a particular runway might require a higher-than-normal glide slope angle.

(2) Under certain wind conditions, unusual terrain features near the airfield can cause turbulence. This could threaten nearby aircraft. Helicopters also can create turbulence that could result in harm to light aircraft.

(3) Prohibited areas, mountains, or other obstacles directly in line with the end of the runway can require the pilot to turn the aircraft abruptly to the right after takeoff (liftoff).

(4) If friendly forces fire either artillery or mortars within the control zone, the pathfinder might need to tell the pilot the origin, range, direction, and maximum ordinate of the firing. He also tells the pilot about any air strikes in the control zone, especially those by high-performance aircraft. He also gives the pilot any available information about the enemy situation.

3-10. SPACING TECHNIQUES

Spacing provides more separation between aircraft in the traffic pattern. This relieves traffic congestion. The pathfinder controller uses two methods to obtain the required separation: the 360-degree turnout and the traffic pattern extension.

a. **360-Degree Turnout**. Except on the final approach, the pathfinder can issue instructions for the 360-degree turnout (a two-minute maneuver) at any point in the traffic pattern. When a pilot receives instructions to begin a 360-degree turnout, he turns away from the center of the landing site, makes a wide circle, and reenters the traffic pattern at about the same point where he left it (Figure 3-2 and Table 3-5 [both on page 3-10]). If the first turnout does not give him enough room, he might have to make more turnouts.

Pathfinder:	DELTA THREE SIX, BEGIN THREE SIX ZERO DEGREE TURNOUT FOR SPACING AND REPORT REENTRY.
Pilot:	*ROGER. [After completing turnout] LIMA ONE SIX, DELTA THREE SIX HAS REENTERED.*
Pathfinder:	DELTA THREE SIX, ROGER, REPORT BASE.
Pilot:	*ROGER.*

b. **Three Legs of Traffic Pattern**. The pathfinder can only extend the traffic pattern on three legs: upwind, crosswind, and downwind (Figure 3-3). He can only extend one leg at a time. He cannot extend the base leg or the final approach, because they run back into the traffic pattern itself. When giving instructions to extend the traffic pattern, the pathfinder includes the length of the extension. The extension normally measures twice the original length of that leg. The pathfinder takes care to ensure that he does not extend the leg so far that he loses visual contact with the aircraft.

3-11. FINAL LANDING INSTRUCTIONS

Final landing instructions consist of a current wind reading (direction and velocity) and clearance to land. The pathfinder includes any change to the situation in the final landing instructions, which he issues as soon as the pilot reports from the designated point. As a rule, once the pathfinder clears an aircraft to land, he can only rescind that clearance in extreme situations.

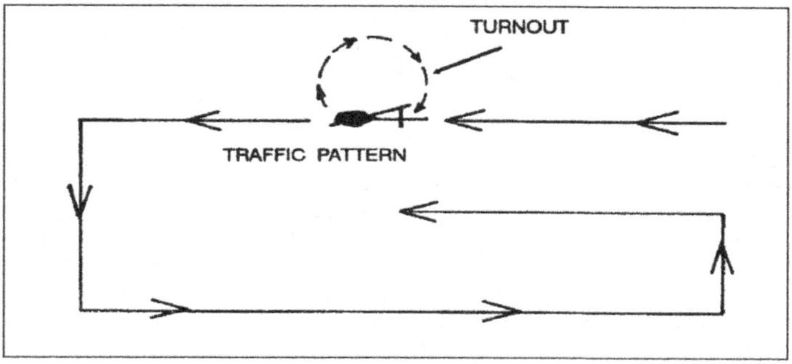

Figure 3-2. A 360-degree turnout.

SITUATION	REPORTING POINT
AIRCRAFT IN TRAFFIC	BASE LEG OF TRAFFIC PATTERN
STRAIGHT-IN APPROACH	FINAL
AIRCRAFT AUTHORIZED TO FORM OWN APPROACH	FINAL

Table 3-5. A 360-degree turnout.

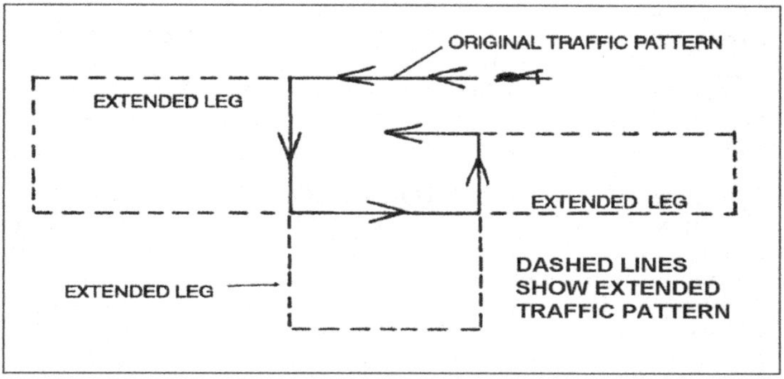

Figure 3-3. Traffic pattern extension.

a. The best reporting points vary with the situation.

b. Aircraft flying in formation, except those flying in trail, and aircraft with a sling load usually form their own approach.

c. If two or more missions arrive at the same time, the controller gives first landing priority to in-flight emergencies, followed by MEDEVAC aircraft. He gives next priority to multiple aircraft and sling-loaded aircraft. Last priority goes to all other flights.

3-12. TAXIING AIRCRAFT

When issuing taxiing instructions, the pathfinder includes a route for the aircraft to follow in the movement area. He also includes instructions for the pilot to hold the aircraft at a specific point, if needed. The pilot moves the aircraft in the loading, maintenance, dispersal, or parking areas without the pathfinder's help but sometimes aided by signalmen.

a. The controller holds a taxiing aircraft short of an active runway by at least two airplane lengths. This ensures that landing aircraft have sufficient clearance.

b. The controller issues concise, easy-to-understand information.

Example: SIERRA TWO SIX, TURN RIGHT AT SIGNALMAN.

TANGO THREE SIX, TURN LEFT AT END OF RUNWAY, OVER.

3-13. MINIMUM AIRCRAFT SEPARATION REQUIREMENTS

During normal operations, pathfinders ensure pilots follow minimum separation criteria. Combat situations may dictate less separation (Figure 3-4, page 3-12).

a. **Arriving Aircraft**. The preceding aircraft (1) taxis off the landing strip before the arriving aircraft (2) crosses the approach end on its final glide (A, Figure 3-4, page 3-12).

b. **Departing Aircraft**. The preceding aircraft either crosses (1) the opposite end of the runway or turns away (2) from the projected path of the departing (3) aircraft before the latter begins its takeoff run (B, Figure 3-4, page 3-12).

c. **Departing and Arriving Aircraft**. The departing aircraft (1) crosses the opposite end of the runway before the arriving aircraft (2) crosses the approach end on its final glide path (C, Figure 3-4, page 3-12).

d. **Departing, Preceding, and Arriving Aircraft**. The preceding aircraft (1) and the arriving aircraft (2) both taxi off the runway before the departing aircraft (3) begins the takeoff run (D, Figure 3-4, page 3-12).

Section III. GROUND-TO-AIR COMMUNICATIONS

Air traffic control requires a rapid and efficient means of communication between aircraft and ground stations. Two-way radio offers the most efficient means, since it allows clear and rapid exchange of information. Not all aircraft have radios that work. A system of visual signals serves as a backup or standby means of communication when the control center or aircraft radio does not work or if the aircraft does not have the control frequency (Table 3-6, page 3-12). Pathfinders may also use colored smoke signals but must coordinate with the aviation unit so that the pilots will know what each color means.

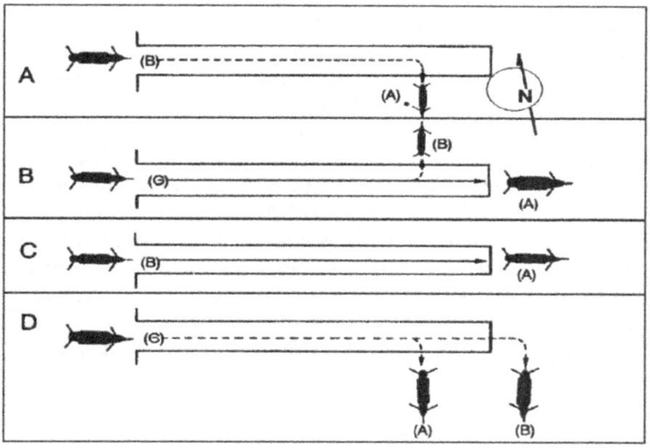

Figure 3-4. Minimum separation requirements.

3-14. ELECTRONIC WARFARE ENVIRONMENT

The pathfinder should expect an active EW environment for all operations. He should make sure he knows the proper ECCM. These include prowords that signal a switch to an alternate radio frequency, transmission authentication procedures, brevity codes, and required reports, when he suspects enemy interference. The pathfinder uses proper radiotelephone procedures and SOI during all operations.

COLOR AND TYPE OF SIGNAL LIGHTS	WHAT THIS MEANS TO AN AIRCRAFT ON THE GROUND	WHAT THIS MEANS TO AN AIRCRAFT IN FLIGHT
Steady green	Cleared for takeoff [or liftoff]	Cleared to land
Flashing green	Cleared to taxi	Return for landing [followed by a steady green light at the proper time]
Steady red	Stop	Give way to other aircraft and continue circling
Flashing red	Taxi clear of landing area of runway in use	Airport unsafe-- DO NOT LAND
Flashing white	Return to starting point [on airfield]	NA
Alternating red and green [general warning signal]	Use extreme caution	Use extreme caution
Red pyrotechnic [red flare]	NA	Do not land for the time being, despite previous instructions.

Table 3-6. ATC light signals on or near LZ.

a. To limit the possibility of compromise, the pathfinder reduces the electronic signature at the LZ or DZ. For this, he depends on thorough mission planning and coordination. He plans control procedures that enable him to execute the mission under radio listening silence. All pathfinder missions seek to achieve this goal.

b. Sometimes the pathfinder has little time to plan the mission. At other times, tactical or meteorological conditions may affect the operation. In either case, the pathfinder may have to use GTA communications to resolve possible conflicts between friendly airspace users and to advise them of previously unknown restrictive landing conditions. These conditions could include wind gusts, hazardous slopes, obstacles, soft landing surfaces, or a limited number of landing points. Training and close liaison with aviation aircrews enable the pathfinder to develop an understanding of what information pertains to the situation. This reduces transmission time.

c. The pathfinder manages any variation due to unknown influences just as he would manage any other exception to set procedure. The landing site is the variable most subject to change. Many conditions could require its relocation. For example, ground fog could cause a delay while the pathfinders move the site to a higher elevation.

(1) Before they know whether a site will support sufficient landing points or an assembly area for the ground unit, the pathfinders must first secure the site and conduct air and ground reconnaissance surveillance.

(2) If the mission is to reinforce or resupply a ground unit in contact, a change in the tactical situation could render the proposed location unsuitable. If the pathfinders locate the proposed site near enemy activity, they will most likely have to move it sometime between planning and execution to ensure that it continues to meet mission requirements.

d. Maintaining radio silence within the LZ is important. Because of this, most air movements require the establishment of a CCP. This ensures a common point from which the pathfinders and the aircraft can refer their relative positions and provide each other time to adjust to other changes.

e. The GTA net is reserved for communications, but the pathfinder cannot assume that all transmissions originate from aircraft. The headquarters in charge of flight plans logs all arrivals so they will know if an aircraft does not arrive at its destination on time. When this happens, they contact intermediate stop points to identify the last known location and to aid in search-and-rescue operations.

3-15. GROUND-TO-AIR TRANSMISSIONS

The pathfinder may encounter endless situations while using GTA. If he can master the following four most common ones, he can handle most situations:

a. **Situation 1--Known Aircraft Location**.

(1) ***Initial Contact***. The pilot radios transmission at coordinated time and location. After establishing two-way communications, the controller may abbreviate call signs. With multiple flights, instructions issued by pathfinder GTA communication should identify the particular situation by including that station's call sign at the beginning of the transmission.

Pilot:	*ALPHA ONE LIMA ONE SIX [A1L16], THIS IS ROMEO TWO BRAVO TWO SEVEN [R2B27], OVER.*
Pathfinder:	ROMEO TWO BRAVO TWO SEVEN, THIS IS ALPHA ONE LIMA ONE SIX, OVER.

Pilot: *THIS IS BRAVO TWO SEVEN, CCP INBOUND, OVER.*

Pathfinder: THIS IS LIMA ONE SIX, STATE TYPE, NUMBER, AND INTENTIONS, OVER.

Pilot: *THIS IS BRAVO TWO SEVEN, FOUR UNIFORM HOTEL SIXTIES [UH-60s], TROOP DROP-OFF AND SLING LOAD, FOR YOUR SITE, OVER.*

Pathfinder: THIS IS LIMA ONE SIX, ROGER, HEADING THREE TWO FIVE [325], THREE THOUSAND [3,000] METERS. LAND THREE TWO FIVE, SIGNAL ON CALL, LAND ECHELON RIGHT, SLING-LOAD AIRCRAFT USE NUMBER FOUR LANDING POINT, CONTINUE APPROACH FOR VISUAL CONTACT, OVER.

(2) ***Air Traffic Control Information***.

Example: HEADING THREE TWO FIVE, [distance] THREE THOUSAND METERS, OVER. LAND THREE TWO FIVE, OVER.

(3) ***Pertinent Information***.

Example: SIGNAL ON CALL [prepare to establish positive visual contact].

FOUR UNIFORM HOTEL SIXTIES [UH-60s] IN ECHELON RIGHT [advises pilot of the size of landing site].

SLING-LOAD POINT ON NUMBER FOUR TOUCHDOWN POINT [night only].

FIELD ELEVATION, FOUR TWO FIVE FEET [actual field elevation].

(4) ***Advisory Information***. See Table 3-7.

TYPE OF ADVISORY	INFORMATION INCLUDED
FLIGHT	The enemy situation, if it presents a threat to the aircraft.
LANDING	Surface conditions on the landing site such as the presence of sand, mud, or blowing snow.
DEPARTURE	Obstacles in the path of aircraft leaving the site [obstacles above the obstacle departure lights].

Table 3-7. Information included in advisories.

(5) ***Aircraft in Sight***.

Pathfinder: BRAVO TWO SEVEN, THIS IS LIMA ONE SIX, I AM AT YOUR TWELVE O'CLOCK, FIVE HUNDRED METERS, IDENTIFY SIGNAL, OVER.

Pilot: *THIS IS BRAVO TWO SEVEN, I IDENTIFY GREEN SMOKE, OVER.*

(a) At night, during specialized activities such as external load drop-off or pickup, or when unsafe surface conditions require pathfinders to mark specific landing points, the flight leader must know all arrangements so he can organize the flight for landing. Pathfinders identify the site by using a light gun to flash a dot-dash sequence.

Pathfinder:	THIS IS LIMA ONE SIX, VISUAL CONTACT [and, once the pilot identifies the site], WIND THREE TWO FIVE AT EIGHT, CLEAR TO LAND, OVER.

(b) Once the pilot identifies the site, the pathfinder issues final landing instructions. If the controller already has other aircraft flying in a traffic pattern, he places the incoming aircraft into the traffic pattern at a safe and convenient location. Then, he instructs the pilot to report base. When the pilot reports base, the pathfinder issues final landing instructions. For special situations, instead of placing the aircraft in the traffic pattern, the controller may tell the pilot to circle left or right. Then the controller will issue final landing instructions.

(6) ***Departure Instructions***. If the departure heading differs from the land heading, the controller gives the departure heading as the first element of the departure instructions.

Pilot:	*LIMA ONE SIX, THIS IS BRAVO TWO SEVEN, READY FOR DEPARTURE, OVER.*
Pathfinder:	THIS IS LIMA ONE SIX, WIND THREE TWO FIVE AT EIGHT, CLEAR TO DEPART, STATE INTENTIONS, REPORT CLEAR OF LANDING ZONE, OVER.
Pilot:	*THIS IS BRAVO TWO SEVEN, RIGHT BREAK, AFTER DEPARTURE, OVER.*
Pathfinder:	THIS IS LIMA ONE SIX, ROGER, OVER.
Pilot:	*THIS IS BRAVO TWO SEVEN, CLEAR TO THE WEST, OVER.*
Pathfinder:	THIS IS ALPHA ONE LIMA ONE SIX, ROGER, OUT.

b. **Situation 2--Aircraft Reporting from a Cardinal Direction and Distance**.

(1) Often, units conduct mutually supporting helicopter operations to increase the security of an LZ operation. For example, a team of observation and attack helicopters may screen the LZ. Because no aircraft plan to land there, and because the utility or lift aircraft know the LZ and screen team's location from communications over internal UHF or VHF radio nets, the screening helicopters need not contact the pathfinder. However, if they learn an aircraft does need to land at the LZ, the screening helicopter team responds differently to initial contact.

(2) Due to a possible conflict with aircraft departing the landing site in the same direction, the pathfinder must track the inbound aircraft's course and advise mission aircraft of the unexpected arrival. To accurately track the aircraft and control the situation, the pathfinder uses a commonly known point in the direction of the aircraft. He can use a prominent terrain feature, a checkpoint, or an aerial control point previously established by the ground unit for maneuver control. This situation matches Situation 1 exactly, except that the controller does not give the aircraft's heading and distance.

c. **Situation 3--Aircraft with In-Flight Emergency**.

(1) An in-flight emergency occurs when an aircraft develops a mechanical problem that challenges the pilot's ability to maintain control. Because the pilot must focus on the problem with the aircraft, the pathfinder helps by moving other air traffic away from the one having the problem, which has first priority. If the emergency develops before initial contact, OPSEC requires a full information exchange, just like in a standard transmission.

(2) After the pilot declares the emergency, the situation continues as follows:

Pilot: *ALPHA ONE LIMA ONE SIX [A1L16], THIS IS CHARLIE ZERO WHISKEY ZERO TWO [COW02], IN-FLIGHT EMERGENCY [MAYDAY], OVER.*

Pathfinder: THIS IS LIMA ONE SIX, WIND ZERO THREE FIVE AT SIX, CLEAR TO LAND, STATE INBOUND HEADING, OVER.

Pilot: *THIS IS WHISKEY ZERO TWO, HEADING TWO SIX ZERO, OVER.*

Pathfinder: ALL STATIONS, THIS IS ALPHA ONE LIMA ONE SIX, BE ADVISED, IN-FLIGHT EMERGENCY APPROACHING FROM THE EAST, REMAIN CLEAR OF LANDING SITE AND MAINTAIN RADIO SILENCE UNTIL EMERGENCY HAS BEEN TERMINATED - BREAK - WHISKEY ZERO TWO, CAN I BE OF FURTHER ASSISTANCE, OVER.

Pilot: *THIS IS WHISKEY ZERO TWO, NEGATIVE, OVER.*

Pathfinder: THIS IS LIMA ONE SIX, ROGER, OVER.

(3) The controller tells the emergency aircraft of any aircraft that remain on the landing site.

Pathfinder: WHISKEY ZERO TWO, BE ADVISED, TWO UNIFORM HOTEL ONES ON NORTH END OF SITE.

(4) Only the pilot who originally declared the emergency can terminate that same emergency. Once the pilot does so, the pathfinder transmits a net call to inform all stations that normal operations can continue.

Pathfinder: ALL STATIONS, THIS IS ALPHA ONE LIMA ONE SIX, EMERGENCY HAS TERMINATED, I CAN ACCEPT TRAFFIC, OVER.

(5) Departure instructions are the same as those given for Situation 1.

d. **Situation 4--Disoriented Aircraft**.

(1) During limited visibility, adverse weather, in-flight emergencies, or when he has no map, a pilot may not know the location of the landing zone. Also, he may not see any easily identifiable land point. In such cases, the pathfinder can help the pilot by directing him either to a known location or to the LZ. At terrain flight altitudes, and in some environments, the pilot may experience disorientation of as little as 200 meters. The pathfinder may hear but not see the aircraft. Pilots who have FM homing equipment onboard may use that to orient themselves during the initial contact, without having to

ask for a long or short count. Because it requires the ground station to increase transmissions, FM homing risks loss of SIGSEC.

(2) In this example, an aircraft at the CCP cannot establish voice communication with the pathfinder due to low altitude or radio interference. Knowing the landing zone location, but unsure of the exact location of the landing site, the pilot continues his flight closer to the center of the zone.

Pilot:	*ALPHA ONE LIMA ONE SIX [A1L16], THIS IS CHARLIE TWO ECHO THREE FOUR [C2E34], OVER.*
Pathfinder:	CHARLIE TWO ECHO THREE FOUR, THIS IS ALPHA ONE LIMA ONE SIX, OVER.
Pilot:	*THIS IS ECHO THREE FOUR, FOUR UNIFORM HOTEL ONES [UH-1s] ARE INBOUND FOR LANDING, REQUEST NAVIGATIONAL ASSISTANCE, OVER.*
Pathfinder:	THIS IS LIMA ONE SIX, DO YOU HAVE FM HOMING CAPABILITY?
Pilot:	*THIS IS ECHO THREE FOUR, AFFIRMATIVE, OVER.*
Pathfinder:	THIS IS LIMA ONE SIX, SHORT COUNT FOLLOWS: 1-2-3-4-5-5-4-3-2-1. END SHORT COUNT, STATE INBOUND HEADING, OVER.
Pilot:	*THIS IS ECHO THREE FOUR, SAY AGAIN, OVER.*
Pathfinder:	THIS IS LIMA ONE SIX, ROGER, ORBIT PRESENT LOCATION, DESCRIBE PROMINENT TERRAIN FEATURES, STATE LAST KNOWN LOCATION, HEADING, AND DISTANCE FLOWN, OVER.
Pilot:	*THIS IS ECHO THREE FOUR, CCP HEADING THREE SIX ZERO, TWO THOUSAND METERS, I SEE A THREE-ACRE POND WITH DAM ON THE SOUTH, ORIENTED EAST-WEST, OVER.*
Pathfinder:	[Plots the course correction and continues with the standard transmission.] THIS IS LIMA ONE SIX, HEADING TWO NINE ZERO, EIGHT HUNDRED METERS, [gives advisories if any], OVER.

(3) The standard ATC information continues as in Situation 1 and ends with--

Pathfinder:	DESCRIBE PROMINENT TERRAIN FEATURES EN ROUTE, OVER.

THIS CHAPTER IMPLEMENTS STANAGs 3281 AND 3619 AND QSTAG 585.

CHAPTER 4
HELICOPTER LANDING ZONES

Helicopter landing zones contain one or more helicopter landing sites. Each landing site has a control center and, in most cases, a manned or unmanned release point (STANAG 3619).

Section I. SELECTION OF LANDING SITES

The ground unit commander coordinates with the supporting aviation unit to select helicopter-landing zones that can support the ground tactical plan.

4-1. REQUIREMENTS

Minimum landing space requirements and minimum distance between helicopters on the ground depend on many factors. If the aviation unit SOP does not spell out these requirements, the aviation unit commander works with the pathfinder leader. The final decision about minimum landing requirements rests with the aviation unit commander. In selecting helicopter-landing sites from maps, aerial photographs, and actual ground or aerial reconnaissance, he considers the following factors:

a. **Number of Helicopters**. To land a large number of helicopters at the same time, the commander might have to provide another landing site(s) nearby. Or, he can land the helicopters at the same site, but in successive lifts.

b. **Landing Formations**. Helicopter pilots should try to match the landing formation to the flight formation. Pilots should have to modify their formations no more than necessary to accommodate the restrictions of a landing site (Figure 4-1, page 4-2). However, in order to land in a restrictive area, they might have to modify their formation somewhat.

c. **Surface Conditions**. Pathfinders choose landing sites that have firm surfaces. This prevents helicopters from bogging down, creating excessive dust, or blowing snow. Rotor wash stirs up any loose dirt, sand, or snow. This can obscure the ground, especially at night. Pathfinders remove these and any other debris from landing points, since airborne debris could damage the rotor blades or turbine engine(s).

d. **Ground Slope**. Pathfinders choose landing sites with relatively level ground. For the helicopter to land safely, the slope should not exceed 7 degrees (Figure 4-2, page 4-3). Whenever possible, pilots should land upslope rather than downslope. All helicopters can land where ground slope measures 7 degrees or less.

(1) ***Observation and Utility Helicopters***. When the slope exceeds 7 degrees, observation and utility helicopters must terminate at a hover to load or off-load personnel or supplies.

(2) ***Large Utility and Cargo Helicopters***. When the slope measures between 7 and 15 degrees, pathfinders advise the pilots of large utility and cargo helicopters.

NOTE: To determine slope in percentage or degrees, express all measurements in either feet or meters, but not both. If the map sheet expresses elevation in meters, multiply by three to convert meters into feet. If the map sheet expresses elevation in feet, divide by three to convert to meters.

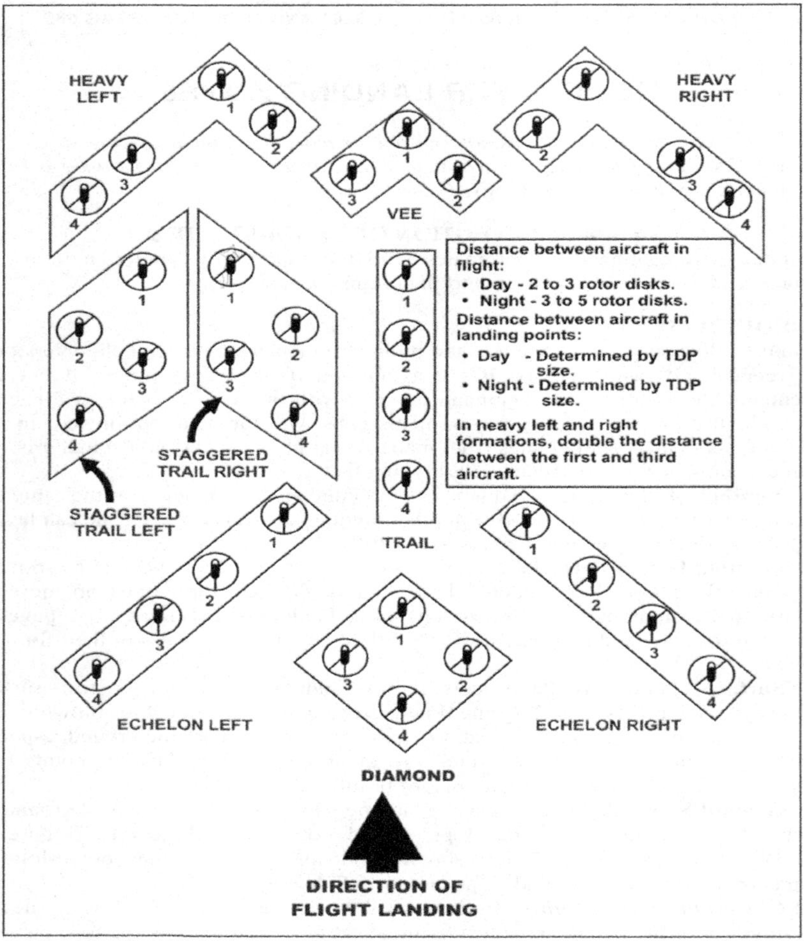

Figure 4-1. Standard flight and landing formations.

e. **Approach and Departure Directions**. To land or to take off, especially at night, the helicopter pilot generally chooses the approach or departure path with the lowest obstacles. Ideally, this allows him to fly into the wind. Depending on the helicopter's capabilities, if only one direction offers a good approach, or to make the most of available landing area, the pilot might be able to land with a crosswind of 0 to 9 knots or a tailwind of 0 to 5 knots. When wind speeds exceed 9 knots, the pilot must land into the wind. The same considerations apply to departures from landing sites.

f. **Prevailing Wind**. Except when the crosswind velocity exceeds 9 knots during a landing, the prevailing wind requires less attention than it does on the approach and departure routes. The wind affects smaller aircraft more than larger, more powerful ones.

GROUND SLOPE EXPRESSED IN DEGREES

The approximate slope angle may be calculated by multiplying the gradient by 57.3. This method is reasonably accurate for slope angles under 20 degrees.

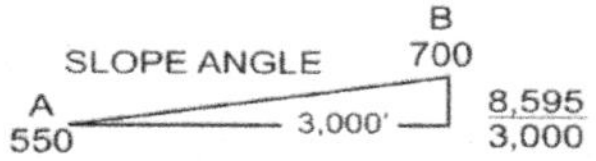

VD = B-A = 150
HD = 3,000

DEGREE OF SLOPE $\frac{150 \times 57.3}{3,000}$

$\frac{8,595}{3,000}$ = APPROXIMATELY 3° OF SLOPE

GROUND SLOPE EXPRESSED AS PERCENTAGE

To determine the percent of ground slope, divide the vertical distance (VD) by the horizontal distance (HD) and multiply by 100.

$$\text{PERCENT of SLOPE} = \frac{VD}{HD} \times 100$$

Verticle distance is the difference in field elevation between the two ends of the landing site. Always round number up to the next whole number.

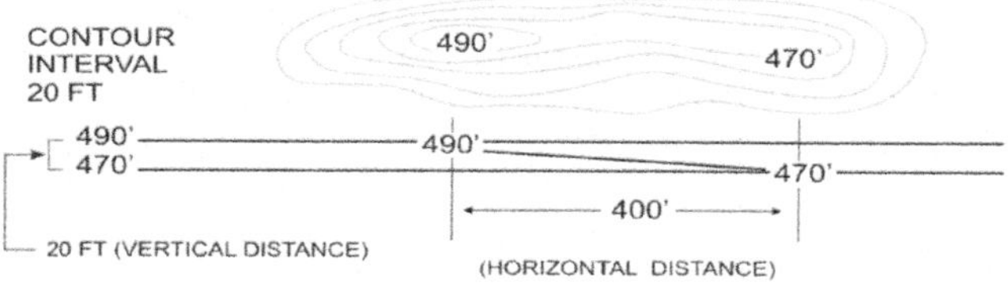

$400 \overline{)20.00}$ = .05   .05 x100 = 05.00 = 5 PERCENT

PATHFINDER SLOPE LANDING RULES

Do not land small utility and observation aircraft on slopes exceeding 7 degrees.
Give large utility and cargo aircraft an advisory if ground-slope is between 7 and 15 degrees.
Always advise pilot when landing wheeled aircraft on a sideslope.

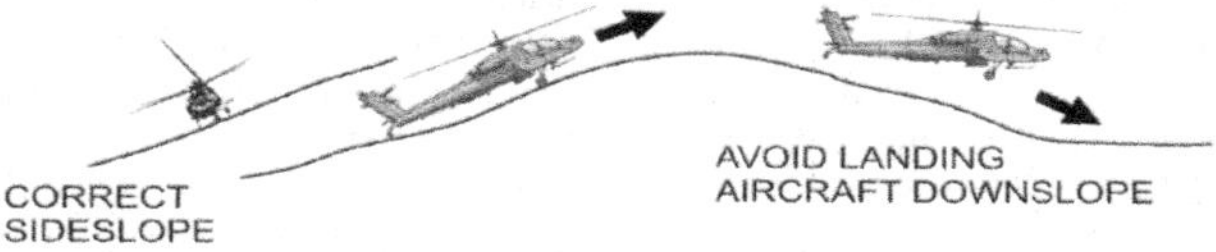

Figure 4-2. Determining ground slope.

g. **Density Altitude**. Altitude, temperature, and humidity determine the density altitude. Planners should try to remember that as the density altitude increases, the size of the LZ also increases. High, hot, humid conditions decrease lift capabilities.

h. **Loads**. When fully loaded, most helicopters can neither climb nor descend vertically. They need a larger area and better approach or departure routes than when they carry lighter loads.

i. **Obstacles**. Landing zones should have no tall trees, power lines, or similar obstructions on the approach and departure ends of the landing site. Pathfinders must remove any obstacles within the landing site. This includes any rocks, stumps, and holes; and thick grass or brush over 0.45 meter (18 inches). Planners figure on an obstacle ratio of 10 to 1 (ten to one). That is, if a helicopter must approach or depart directly over a 10-foot tall tree, then the landing point must have 100 feet of horizontal clearance.

NOTE: The helicopter unit commander makes the final decision on minimum landing requirements. He bases his decision on the effects of air density, slope, and surface conditions. He explains these requirements verbally during early mission planning.

4-2. ALTERNATE SITES

Enemy action, unfavorable terrain, or changes in the tactical or logistical situation can require alternate landing sites. The ground unit commander usually selects these to support the tactical plan. He (or his representative) decides when to use them based on the recommendations of the aviation unit commander and the pathfinder on the site. The commander uses the fastest means to get instructions for using alternate sites to the pathfinders. Neither pathfinder nor aviation unit commanders can shift to an alternate LZ(s) unless the supported ground unit commander has delegated that authority to them.

4-3. LANDING POINTS

The pathfinder leader designates areas or points that can support the weight of a helicopter. A helicopter requires a level, clear, circular area that measures between 25 and 100 meters in diameter. The type of helicopter determines the exact size of the area. The ground unit and pathfinders work together to clear from the landing point area all trees, brush, stumps, or other obstacles that could damage the helicopter. Usually, a helicopter requires a larger area for landing in the dark than in daylight. The size of the landing area also depends on the type of helicopter, the nature of the load, the climate, and visibility.

a. **Choose a Hard Surface**. Pathfinders choose a landing point with a "hard surface." That is, the surface of the landing point must allow a fully loaded helicopter to land, restart, and leave again, all without sinking into the ground.

b. **Clear to Ground Level**. Pathfinders must clear the entire landing point of any loose material that the rotors could blow up. The term is "cleared to ground level." Unless a fire risk exists, they need not clear grass less than 0.3 meter (1 foot) high, as long as the field is level. They can cut down on dust by wetting down dry dirt. They can reduce snow to reveal hazards, then pack it down firm, which will also reduce the amount blowing around.

c. **Clear Around Obstacles**. Ground troops must do everything they can to improve landing point surfaces so aircraft can land. However, even if pathfinders cannot clear ground obstructions, they can perform some helicopter operations without the helicopter landing. However, they must still clear and mark the area just as they would if the helicopter were going to land. Helicopters hover above the ground obstructions that prevent them from landing.

d. **Measure the Shortest Distances Between Landing Points**. In a landing site, pathfinders measure the minimum distances between landing points, from center to center. When aircraft sizes vary, pathfinders separate landing points by the most generous measure, allowing 100 meters from center to center of the landing points (Figure 4-3, page 4-6). At the least, these distances must measure--

(1) Size 1 Landing Point. 25 meters.
(2) Size 2 Landing Point. 35 meters.
(3) Size 3 Landing Point. 50 meters.
(4) Size 4 Landing Point. 80 meters.
(5) Size 5 Landing Point. 100 meters.

Section II. ORGANIZATION AND DUTIES

The commander task organizes the pathfinder element to set up and operate the installations required by the supported unit's tactical plan. They may set these up within a single LZ or separate them widely throughout a large AO. The pathfinder leader normally stays at the most important site. To set up and operate one helicopter LZ, the commander task organizes the pathfinder element into two working parties-a reconnaissance party and a marking party. Each site requires its own landing site party. The control center party and the release point party provide the same function for LZs or DZs.

4-4. CONTROL CENTER

The CC coordinates aircraft in and around an LZ or DZ and promotes a safe, orderly, and speedy flow of air traffic. Upon arrival in the area, the pathfinder leader selects the exact location of the CC. He positions it to allow visual control of aircraft in and around the LZ or DZ.

a. For helicopter LZs, the most desirable CC location is along the aircraft flight route, but displaced from the landing site. This helps prevent enemy EW assets from compromising the actual landing site location, even if the tactical situation dictates that the pathfinder leader remain on the site for control purposes. For an LZ with more than one landing site, or for any LZ during reduced visibility, the pathfinder leader locates the CC where it can act as a manned RP or final approach fix to provide positive navigational assistance to arriving aircraft.

(1) The RP is an established traffic control checkpoint. It is the final navigational checkpoint for aircraft approaching the landing site or approaching air-delivery facilities in an LZ or DZ.

(2) During the air movement phase of an air assault operation, helicopter serials also use the RP as a final coordination point for control of planned ground or aerial supporting fires in and around LZs. The air movement commander staffs the RP only when he expects tough navigational problems. He tentatively chooses its location from maps or from aerial photographic studies. He looks for an easily identifiable point on the planned flight route to the landing site. He looks for a location that will take advantage of long-range electronic and visual navigation aids.

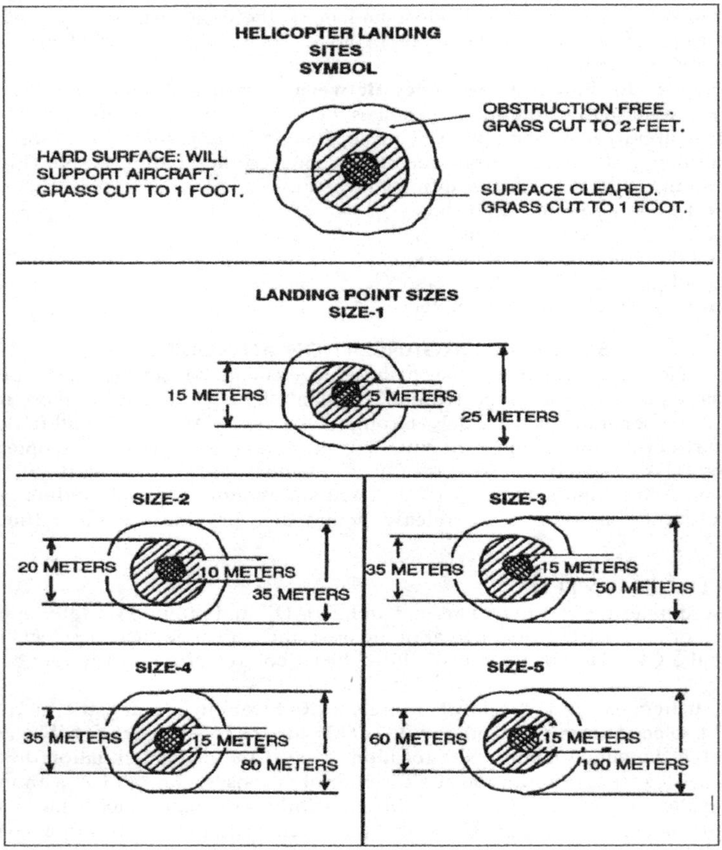

Figure 4-3. Landing points.

b. For single helicopter landing sites within a single LZ, the site itself offers the best location for GTA communication. Especially at night, positioning here allows the pathfinder ATC to observe the final approach of helicopter formations. It helps him make sure pilots align correctly on the required landing direction. It also helps him ensure that they clear any obstacles.

c. The pathfinder leader organizes the control center to meet mission requirements. The control center can consist of a single pathfinder. This soldier can operate the GTA radio for a limited period at a small site, or the control center can consist of the following staff:

(1) ***LZ or DZ Commander***. He supervises aircraft landings and departures, airdrops, and other pathfinder activities in the LZ or DZ. He may also serve as the GTA radio operator.

(2) ***GTA Radio Operator***. He operates the radio used to maintain communications with pilots. He also provides advisories for his airspace, as needed.

(3) ***Internal Net Recorder***. Some situations require pathfinders to set up an internal net to communicate with other pathfinder elements. An internal net recorder (INR) runs this net. The INR can help control aircraft by observing them and recording their arrivals, departures, and loads. He records air traffic to and from the helicopter-landing zone (HLZ) on DA Form 7461-R (Figure 4-4, page 4-8). This form serves as a log or manifest. If an aircraft fails to arrive at its destination, search and medical units can check the applicable DA Form 7461-R and focus their search between the last departing station and the gaining station. The recorder completes the form as follows:

(a) *PFDR Det*. Write the name of the pathfinder detachment operating the LZ.
(b) *Supported Unit*. Write the name of the supported unit.
(c) *Period (DTG)*. Write the date and time of the mission.
(d) *Operation*. Write the name of the mission.
(e) *Designation*. Write the name and location of the site.
(f) *Recorder*. Write your name.
(g) *No. A/C*. Write the number of aircraft in the formation.
(h) *Type A/C*. Write the nomenclature of each type of aircraft in the formation.
(i) *Contact Time*. Write the time of the initial contact with the flight commander.
(j) *Call Sign*. Write the flight commander's call sign.
(k) *Time, Arr*. Write what time the aircraft or formation inserted.
(l) *Time, Dep*. Write what time the aircraft or formation extracted.
(m) *Load Type, Ins*. Write what type of load the aircraft inserted.
(n) *Load Type, Ext*. Write what type of load the aircraft extracted.
(o) *Destination*. Write the name of the aircraft's or formation's destination on leaving.
(p) *Remarks*. Write anything else here that you think you need to record.

4-5. LANDING SITE PARTY

The landing site party consists of a site team leader and other pathfinders and attached personnel, as required. A single pathfinder may establish and operate a small landing site for a limited time.

a. **Site Team Leader**. The site team leader reconnoiters, establishes, and operates the landing site. He supervises it and, at any time, may supervise the GTA radio operator. Some of his responsibilities include--

(1) Organizing at an objective rally point.
(2) Reconnoitering to determine--

- Long axis.
- Usable area.
- Ground slope (compute).
- Land heading.
- Best landing formation.

(3) Designating sling-load point(s).
(4) Emplacing and briefing the GTA radio operator.

INTERNAL NET RECORD

PFDR Det 1-507 PFDR SECTION 2 — OPERATION "ROYAL COVER"

Supported Unit 1-507 PATHFINDER SCHOOL — DESIGNATION LEDO LZ (YC594062)

Period (DTG) 151430ZJUN00-161830ZJUN00 — RECORDER SSG ALDER

NO. A/C	TYPE A/C	CONTACT TIME	CALL SIGN	TIME		LOAD TYPE		DESTINATION	REMARKS
				APR	DEP	INS	EXT		
5	UH-60	1120	RAVEN 22	1127	1145	PERS 1 M998	–	LAWSON ARMY AFLD	1 SLINGLOAD A/C

DA Form XXX-A-R

Figure 4-4. Example completed DA Form 7461-R.

(5) Clearing touchdown and sling-load points.
(6) Organizing personnel and loads for air movement.
(7) Clearing or marking obstacles.
(8) Preparing for night and day missions.
(9) Continuing to improve the site.

b. **Extra Pathfinders**. These soldiers operate the GTA radio and the pathfinder internal radio net (if established), position and operate navigation and assembly aids, and clear or mark obstacles. Three factors dictate the number of extra pathfinders employed:

- The size of the landing site.
- The expected density of air traffic.
- The number and type of visual and electronic aids used.
- The tactical situation.

c. **Commander**. The commander can attach other soldiers from supported units to the landing site party. The pathfinders brief and rehearse attached soldiers. Only pathfinders reconnoiter actual landing areas, but attached personnel can--

- Reconnoiter other areas.
- Provide security.
- Help pathfinders set up and operate the landing site.
- Reconnoiter and mark assembly areas.
- Operate assembly aids.

Section III. LANDING SITE OPERATIONS

Before they can start using a landing site, pathfinders need only pick its location and set up communications in it. They continue marking and improving the site continually, until it can support the ground tactical plan.

4-6. COMMUNICATIONS

As soon as they arrive at the landing site, pathfinders set up communications in the GTA net. If needed, they also set up the pathfinder internal net. They monitor these radio nets continuously, unless directed otherwise, until they complete operations at the site.

a. Tactical situation permitting, pathfinders locate each helicopter landing site within ground communication range of the other sites and manned RPs. The range of available radios dictates whether facilities within the LZ can communicate with each other.

b. The commander of the landing site for utility and cargo helicopters quickly reconnoiters the area to determine the exact direction of landing. He calculates an intercept heading from the RP, if necessary. He selects the location of the landing point of the lead helicopter of each flight. Then, he decides if the terrain or situation dictates any change to the planned landing formation. The site commander has pathfinders or other personnel compile landing instructions for transmittal to inbound helicopters. He also has them remove or mark obstacles in or around the site.

4-7. FLIGHT FORMATIONS

Ideally, all helicopters land at the same time in a planned flight formation. The landing site commander includes this information in his landing instructions to the flight leader. Pathfinders lay out the landing site in a location where helicopters will not fly directly over aircraft on the ground. The layout of the site also depends on the landing space available, the number and type of obstacles, unit SOPs, and prearranged flight formations.

4-8. LANDING ZONE AND OBSTACLE MARKING

For daylight operations, pathfinders use only smoke, or some other minimal identification means to mark LZs. For night operations, they use lanterns, field expedients, or both to show the direction of landing and to mark individual landing points (Figures 4-5 through 4-9, pages 4-11 through 4-15; and Figure 4-10, page 4-16). For daylight and night air assault operations, they mark all obstacles. (Section V provides more detailed information about conducting night operations.)

a. At night, pathfinders can use lights of different colors (except red, which marks obstacles) to designate different helicopter sites or to separate flights within a larger formation. A lighted "T" or inverted "Y" indicates both the landing point for the lead helicopter of each flight and the direction of approach (Figure 4-11, page 4-17). Other lights mark touchdown points for the other helicopters in the flight. Each helicopter should land with its right landing gear or its right skid 5 meters left of the light.

b. For security, pathfinders and the ground unit cover or turn all lights upside down until the last practical moment before a helicopter arrives. Then, they beam the lights in the direction from which the lead helicopter is approaching, and a signalman directs its landing.

NOTE: Because the infrared lights could be too bright for the aircrew member's NVG, he might have to look under it in order to distinguish the colors.

4-9. AIR ASSAULTS

During daylight air assault operations, pathfinders use red-colored panels or other red, easily identifiable means to mark any hard to detect, impossible to remove obstacles such as wires, holes, stumps, and rocks. During nighttime air assault operations, pathfinders use red lights to mark any obstacles within the landing site that they cannot reduce or remove.

a. In most combat situations, the need for security keeps pathfinders from using red lights to mark treetops on the departure end of a landing zone. However, in training or in a rear area landing site, they do use red lights. If they cannot mark obstacles or hazards, they must fully advise aviators of existing conditions by GTA radio. In any case, the pathfinder landing site leader makes sure that pathfinders mark the most dangerous obstacles first and, if possible, that they remove them.

b. If required to do so by the supported unit, pathfinders can mark initial assembly points for troops, equipment, and supplies. They should choose locations that help ensure the quick, efficient assembly and clearing of the helicopter site. If the unit will use the assembly areas, the ground unit commander selects their locations. If needed, supported ground unit soldiers go with the pathfinders to reconnoiter and mark the unit assembly areas, set up assembly aids, act as guides, and help with landing and unloading operations. Having this help ensures that the pathfinders can rapidly clear troops, supplies, and equipment from the landing points.

c. Pathfinders have a limited capability to secure a landing site. If they precede the initial assault elements into a landing site, soldiers from the supported ground unit could go with them for security.

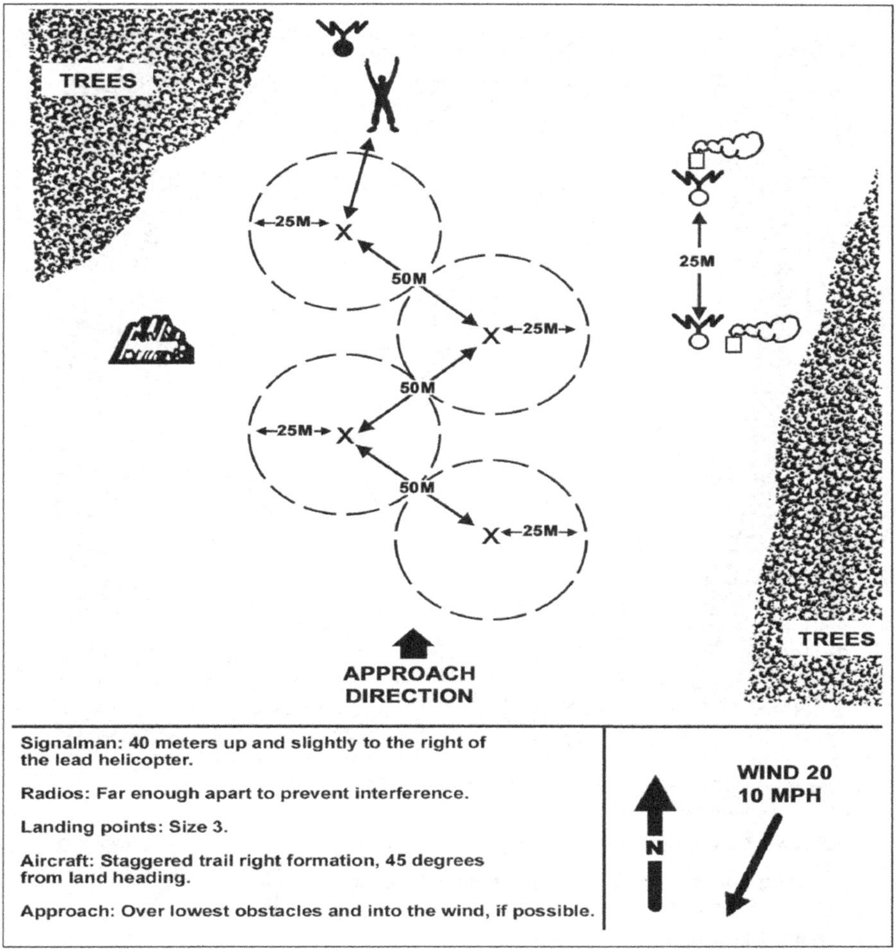

Figure 4-5. Helicopter day landing site, staggered trail-right formation.

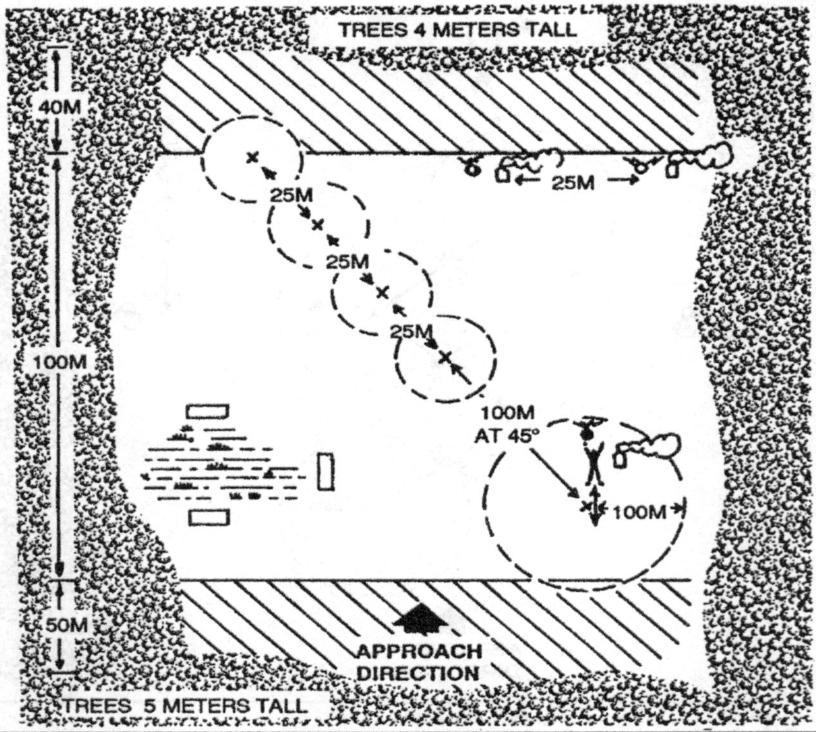

Signalman is 20 meters up and slightly to the right of cargo helicopter .

Radios are far enough apart to prevent interference.

All landing points are size 1 except cargo helicopter point which is size 5.

Aircraft are in echelon right formation.

Aircraft are 45° from land heading.

Ground markings are secured.

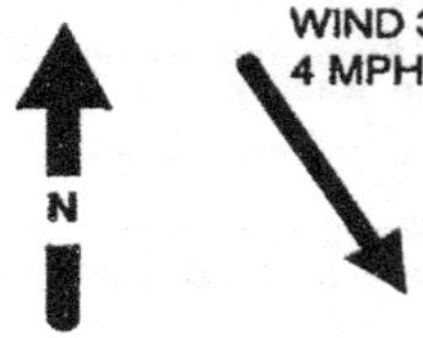

Figure 4-6. Helicopter day landing site, echelon right formation.

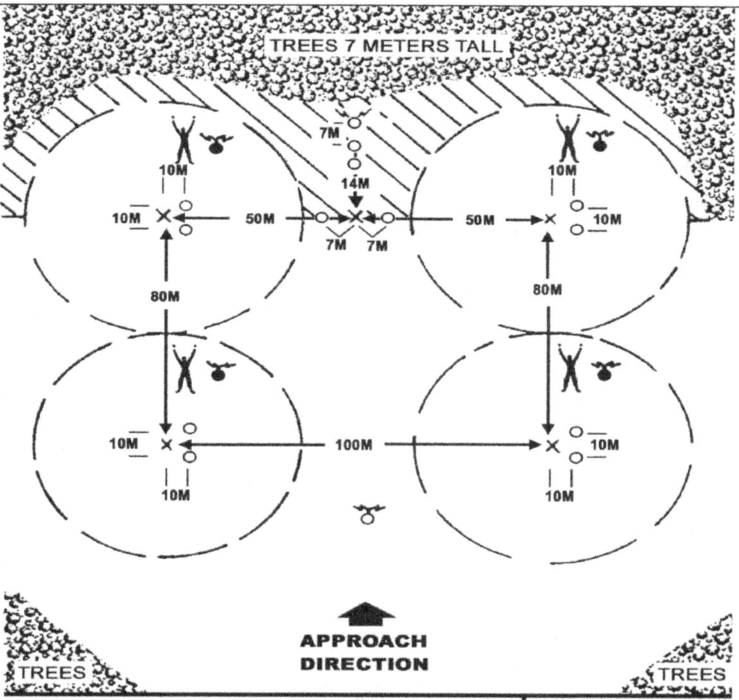

Signalmen are 20 meters up and slightly to the right of each landing point.

All landing points are size 4, using inverted "Y."

Number of points determined by available area and mission requirements.

Each aircraft approaches inverted "Y," then directed to designated landing point by CC.

Load-point hookup men provided by lifted unit.

Pathfinder internal radio net used throughout site as required.

When triangular reference light are used, signalman is not required.

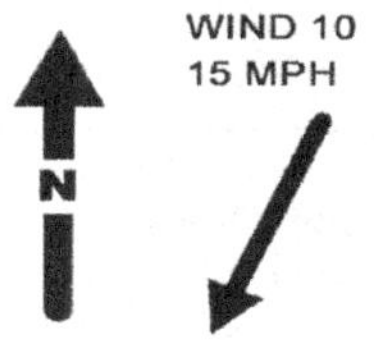

Figure 4-7. Day or night sling-load operation site.

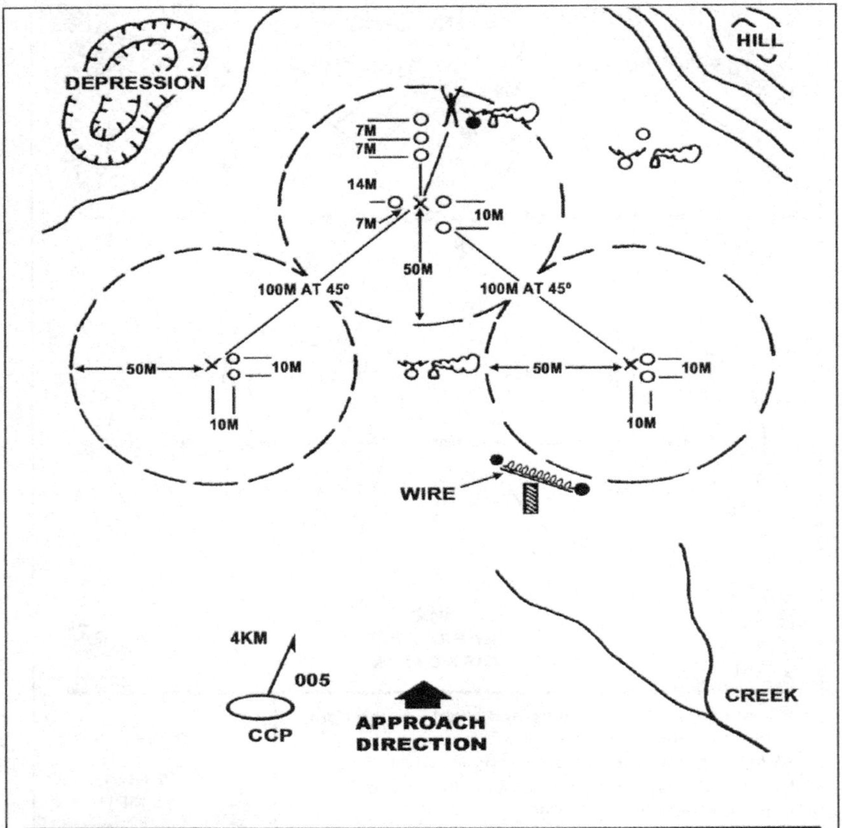

Signalman is 40 meters up and slightly to the right of lead helicopter.
All landing points are size 5.
Aircraft are in vee formation on inverted "Y."
Light added to inverted "Y" for cargo aircraft landing.
Fifth light has additional location (as an additional stem light placed 7M above top stem light).

Figure 4-8. Day or night cargo landing site, "V" formation.

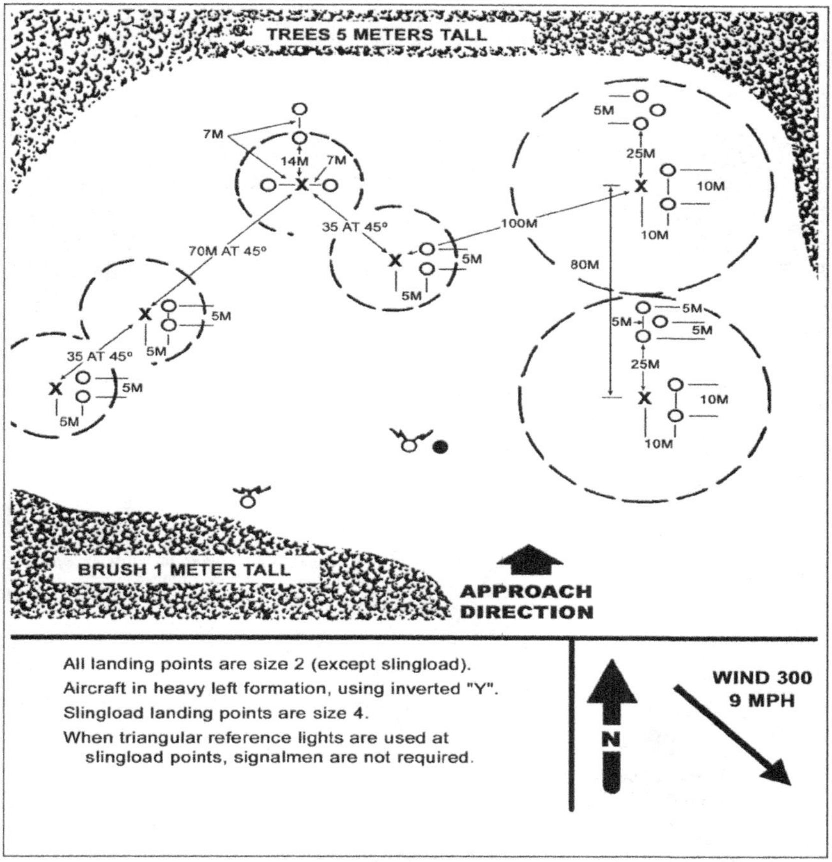

Figure 4-9. Night landing site with landing points for aircraft and sling loads.

4-10. INTERCEPT HEADING

The heading from the RP (or from CCP if the pathfinders do not use an RP) to the landing site coincides as closely as possible with the landing direction to keep the helicopter from having to turn sharply. The larger the formation, the more important this becomes. If a pilot cannot approach the landing site straight on, pathfinders will set up an intercept heading (Figure 4-12, page 4-18). They choose an intercept point far enough from touchdown to allow helicopters in formation a final approach of at least 1 to 2 miles. Flight leaders may need visual steering commands, time and distance information, terrain features, and electronic or visual navigation aids to help them determine the intercept point and the landing direction at the landing site.

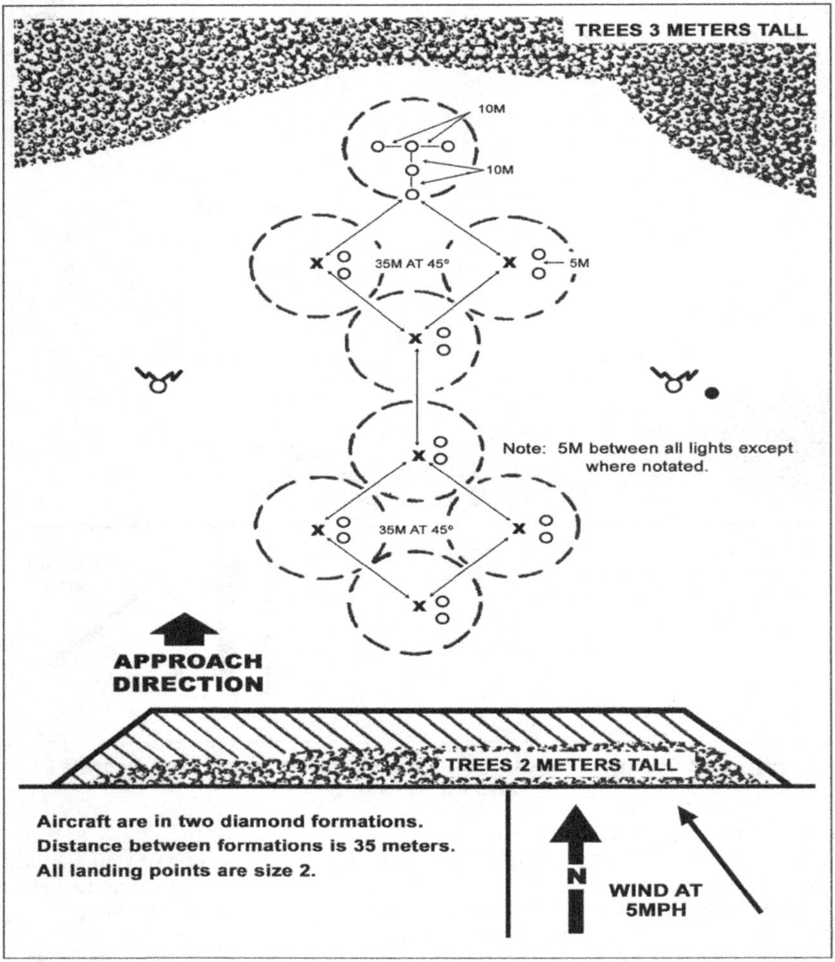

Figure 4-10. Utility helicopter night landing site, diamond formations.

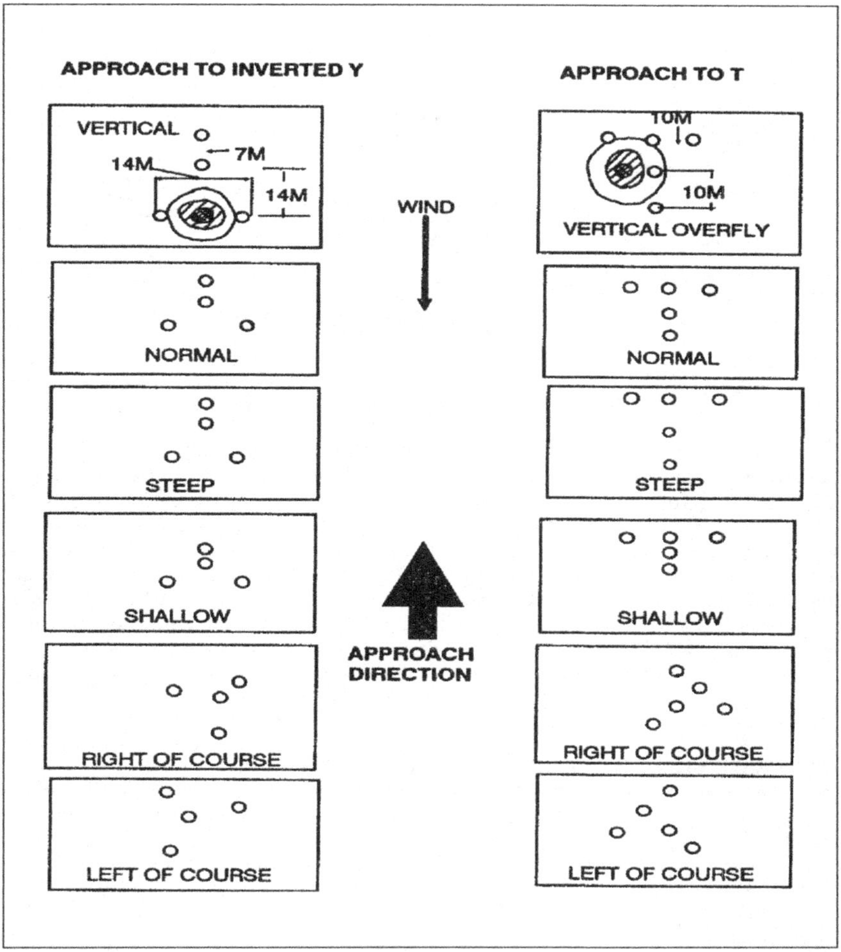

Figure 4-11. Lighted night landing symbols as the pilot would see them from different approach angles.

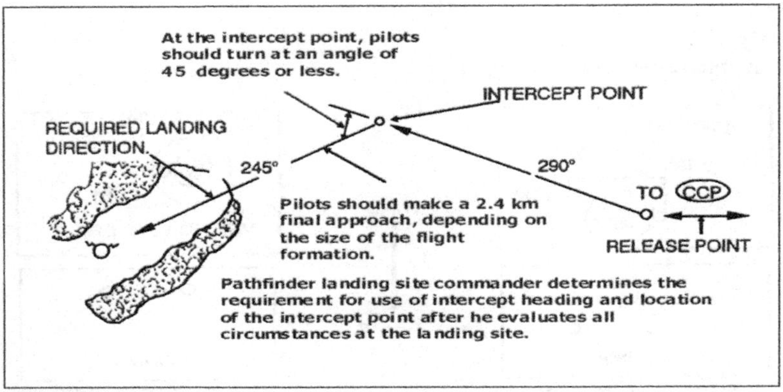

Figure 4-12. Intercept heading technique.

Section IV. LANDING ZONE OPERATIONS

Helicopters approach the LZ along a designated flight route. They normally travel in serials containing four or five helicopters, but they sometimes travel as platoon-sized lifts. One serial may contain a flight for each helicopter site. Flights of medium or heavy transport helicopters (CH-47), carrying artillery or other bulk cargo often arrive at LZs one or two helicopters at a time (Figure 4-13, page 4-19). Later flights follow at the smallest time intervals. These intervals depend on the number of helicopters in each flight, the configuration and conditions of the landing site, and the nature of the cargo to be loaded or unloaded. During planning, the aviation unit commander determines the time between successive flights. Once an operation starts, pathfinders at the site recommend any changes needed to ensure helicopter safety or expedite operations. Night operations often require more time and distance between formations.

4-11. COMMUNICATIONS CHECKPOINT

As each helicopter serial reaches the communications checkpoint (CCP) on the flight route, the flight leader contacts the appropriate helicopter landing site control center.

a. The CC then gives the flight leader the heading from the CCP to the landing site, the landing direction, and other relevant and important information, as follows:

- Enemy situation.
- Friendly fires.
- Field elevation.
- Landing formation.
- Terrain conditions.
- Traffic situation.
- Obstacles.
- Availability of smoke or light gun.
- Next reporting point.

Figure 4-13. Flight techniques.

b. Before reaching the CCP, IAW instructions from the flight leader, all helicopters in a flight switch to the pathfinder control frequency.

NOTE: Pathfinders must stay prepared to provide ATC and navigational aid to all aircraft in and around the landing site, in case those aircraft have no specified flight plan.

c. The helicopter formation continues along the flight route to the RP. The electronic and visual aids at the RP (if manned) help pilots navigate. As each helicopter passes over or near the RP, its flight serial leader reports this to the respective landing site CC. Then the helicopter flies directly to the assigned landing site. The CC at the individual landing site uses visual signals, steering commands, or electronic homing techniques to help any flight that cannot find its landing site.

(1) ***Day Operation Signals***. For daylight operations, you can use different smoke colors for each landing site. You can use the same color more than once, just spread them out. Use smoke only if you have to, because the enemy can see it, too. Try to use it only when the pilot asks for help locating his helicopter site.

(2) ***Night Operation Signals***. For night operations, use pyrotechnics or other visual signals in lieu of smoke. As in daylight, red signals mean "DO NOT LAND," but you can also use it to indicate other emergency conditions. All concerned must plan and know emergency codes. Each flight lands at the assigned site according to CC messages and the visual aids displayed. You can use arm-and-hand signals to help control the landing, hovering, and parking of helicopters.

4-12. AIR CONTROL POINTS

Pathfinders might have to manage ACPs in order to help aircraft en route to the LZ.

a. The ACP party consists of two or three pathfinders, or at least one pathfinder and assistants. The party positions and operates electronic or visual navigation aids, or both.

The party also operate radios in the pathfinder internal net (if used) and the GTA net. The ACP party monitors the GTA net so they can respond at once to any pilot's request for help finding an ACP.

b. The ACP party installs navigation aids as soon as it arrives at the site or as planned. They try to set up all of the aids at the same time. However, if they cannot do this because they have too few people, or for some other reason, then they set them up in the following order:

(1) ***GTA Radio***. The party sets this up first. Then, if the aviation unit commander has asked them to do so, they install the electronic homing beacon. This beacon allows the party to offer long-range guidance. If they do use the beacon, the party sets it up far enough away to prevent excessive radio interference. This also helps keep the enemy from destroying the radios and the beacon at the same time.

(2) ***Visual Navigation Aids***. These navigation aids vary in number and type, depending on aviation unit SOPs and requirements and on the need for security. The ACP party removes any grass or brush that masks their usage of these aids, but they also plan a way to conceal the markings in case they sight enemy aircraft.

(3) ***Internal Net Recorder***. The pathfinder internal net recorder sets up communications with the landing zone CCs as fast as he can. He immediately reports the state of ACP readiness and any information about the local enemy situation, if any. Unless directed to operate a beacon on a definite time schedule, he constantly monitors the radio.

(4) ***Security Personnel***. The ACP party can include attached personnel from the supported units. These personnel provide security. They both move to their assigned locations and take up security positions, or they help set up and operate navigation aids and communications equipment.

Section V. NIGHT OPERATIONS

Daytime visual references (checkpoints for positive identification) are difficult to see at night. Visual aids for night navigation emit illumination. Having too few visual references can cause pilots to concentrate on a single light or group of lights in a concentrated area. This can cause visual illusions, which can then cause vertigo. To prevent this hazardous situation, pathfinders mark LZs with multiple lights and mark landing areas with two or more widely separated lights.

4-13. TACTICAL LANDING LIGHTS

The tactical landing light system provides visual cues for landing in a tactical landing site.

a. When the aircraft approaches from terrain flight altitudes, it should use the inverted "Y" system. Aircraft normally approach a tactical landing site without the aid of the search landing light. The lighting for a tactical LZ can consist of handheld flashlights or "beanbag" lights arranged on the ground.

b. Regardless of the type lighting device used, pathfinders identify the touchdown point with at least two lights. At night, they can designate different helicopter sites with lights of different colors. They may also use different colors to separate flights within a larger formation. A lighted (inverted) "Y" indicates the landing point of the lead helicopter (Figure 4-14, page 4-22).

c. At other touchdown points, helicopters land with the right landing gear or skid just to the left of the light (Figure 4-15, page 4-22). Until the last practical moment, pathfinders hood or turn upside down all lights for security. They also place a signalman at a sling-load point. Then, they beam the lights in the direction from which the helicopters approach.

(1) Pathfinders display an inverted "Y" for cargo aircraft. This marker consists of five lights. Pathfinders place the fifth light IAW prior coordination with the supporting aviation unit (Figure 4-16, page 4-22). The fifth light can go 7 meters from last stem light or 10 meters below right flank light.

(2) Pathfinders will display a NATO landing "T" if an aircraft approaches the LZ from 500 AGL or above, or anytime the pathfinders coordinate in advance with the supporting aviation unit.

(3) Noncargo aircraft require a 5-meter separation between touchdown point and lights, with a 5-meter separation between lights.

(4) Cargo aircraft require a 10-meter separation between touchdown point and lights, with a 10-meter separation between the lights.

d. During darkness, helicopters approach slightly steeper and slower than they would in daylight.

e. Vehicle headlights offer one kind of emergency night lighting. Pathfinders place two vehicles about 35 meters apart and 35 meters downwind of the landing point. They shine their headlights so that their beams intersect at the center of the landing point (Figure 4-17, page 4-22). The helicopter approaches into the wind, passes between the vehicles, and lands in the lighted area. This method does not work well for large helicopters.

CAUTION

When fully adapted to the night, the eyes grow extremely sensitive to any light. Sudden exposure to a light source causes partial to complete loss of night vision.

Thus, take care to avoid exposing pilots to light sources. Also, if pilots are using NVG, avoid shining a light directly at the aircraft, or use light sources compatible with the NVG.

4-14. EXTERNAL LOADS

Employing external loads presents a challenge in the dark. Even so, the pathfinder can use one of several methods. If he lacks sufficient signalmen, he marks the load by placing three reference lights 25 meters in front of the load. He spaces them in a triangle, 5 meters apart. This helps the flight crew during hookup, liftoff, and landing. On liftoff, the aircraft climbs vertically until the load clears the ground. As the helicopter begins to move forward, the pilot applies enough power to maintain a climb that allows the sling load to clear any obstacles on the liftoff path. The shorter the sling, the less altitude required to clear obstacles. To compute the distance required for departure clearance, the pathfinder adds sling length to obstacle height.

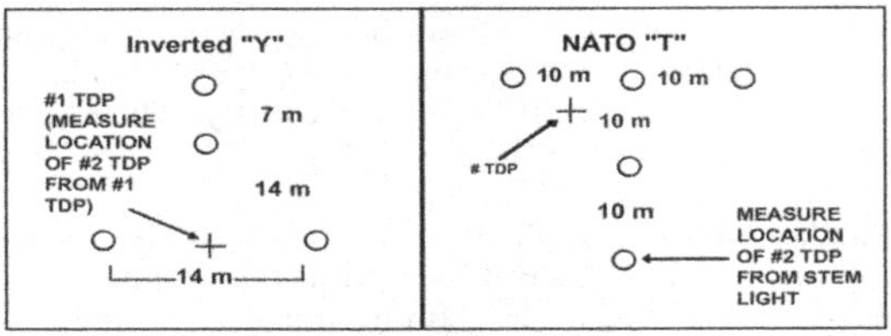

Figure 4-14. Placement of the Inverted "Y" or NATO "T" at the Number One Touchdown Point.

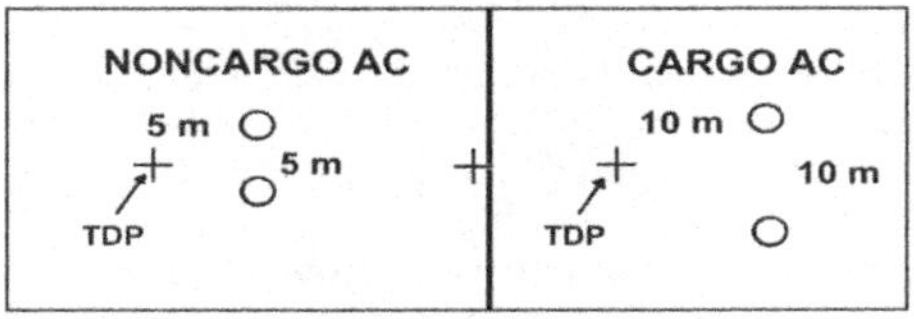

Figure 4-15. Placement of additional touchdown point markings for night use.

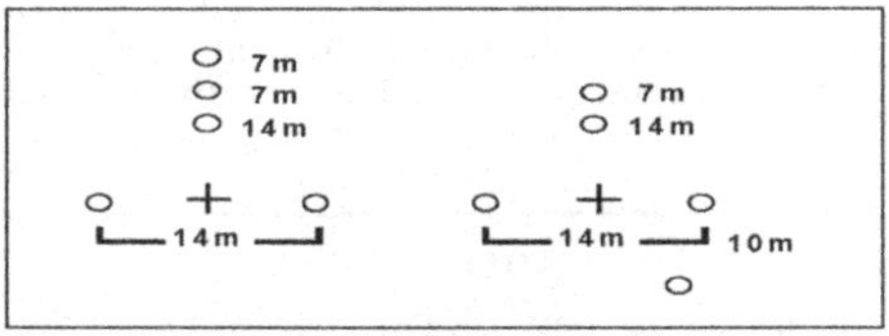

Figure 4-16. Placement of fifth light using inverted "Y," when coordinated.

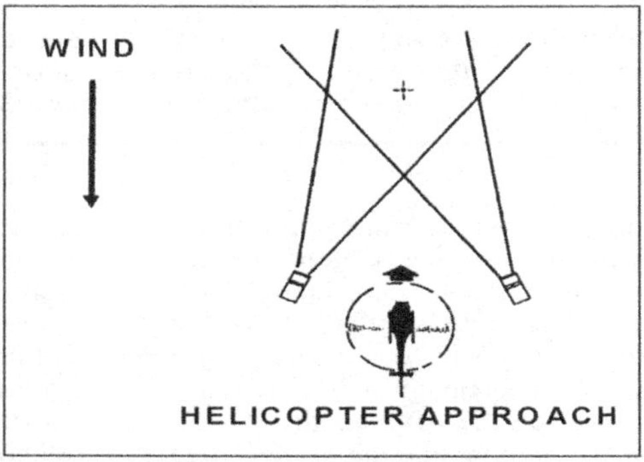

Figure 4-17. Emergency night lighting by vehicle headlights.

4-15. MULTIHELICOPTER OPERATIONS

Only by using NVG can pilots fly safely in formation in a complete blackout and at terrain flight altitudes.

4-16. NIGHT VISION GOGGLES

To operate at terrain flight altitudes during low- or mid-light levels, pathfinders use night vision goggles. Because the lights in the tactical lighting set are too bright for these goggles, pathfinders must place a filter over the clear lens cover. If they do not have filters, they can paint the lens covers or cover them with plastic tape to reduce the intensity of the light.

> **WARNING**
>
> **When your unit trains with or employs the tactical light set, wear a filter over your night vision goggles to prevent eye injury. If you do not have a filter, paint the lens cover or cover it with plastic tape to reduce light intensity.**

Section VI. ENVIRONMENTAL CONSIDERATIONS

The pathfinder unit can expect to support the aviation commander and ground unit commander in many climates and types of terrain. The requirements for establishing a landing site or zone are similar. For aircraft to land safely and quickly in challenging environments, pathfinders must choose and prepare LZs carefully.

4-17. PILOT INPUT

The pilot considers his experiences and his responsibilities to the crew and aircraft before determining whether a proposed landing site is safe.

a. Challenging climatic and terrain environments include extreme hot and cold weather and jungle, desert, and mountainous terrain. (For more detailed information on the climate, terrain, and operational aspects of these areas, see FM 90-3, FM 90-5, and FM 3-97.6 [90-6].)

b. Each area requires the pilot to know and follow special procedures. The pathfinder who also knows these procedures can better advise and assist aviators and the supported ground unit.

4-18. COLD WEATHER

Many parts of the world experience cold weather. Extreme cold and blowing snow pose special problems in ground operations and flight. Pathfinder mission planning includes considering the problems presented by ice, snow, or rain. The pathfinder's knowledge of flight procedures helps him advise the pilot about the existing surface conditions.

a. **Communications**. Most locations allow generally good radio communications. However, atmospheric electricity, such as the aurora borealis, can disrupt them. These events could disturb or block some frequencies. Mountainous terrain also restricts communications. Pathfinders may need to set up relay stations. "Radio skipping" happens often in cold weather areas. RATELOs often hear long-distance radio traffic on tactical FM networks.

b. **Navigation**. In snow-covered areas with flat terrain, pilots may need marked and manned RPs. When aircraft fly over loose snow, the air movement lifts the snow and circulates it into a snow cloud. This often produces a zero-visibility condition known as "whiteout," through which the pilot must take off or land blind (Figure 4-18, page 4-26). Whiteout conditions place extra demands on the landing site party.

c. **Surface Conditions**. The pathfinder evaluates the surface of the ground to see whether aircraft can land without sinking too deep into the snow. He can use a tactical vehicle to test the hardness of the surface. The landing site party may also try to determine the degree of ground slope and whether or not obstacles lie under the cover of snow at each landing point.

(1) ***Distance Between Aircraft***. Pathfinders increase the distance between aircraft to 100 meters and the size of the landing point to 100 meters in diameter.

(2) ***Landing Point Markings***. Marking the landing points presents other problems. A pilot's depth perception is impaired in snow-covered areas. A signalman on the ground provides a useful reference for estimating height. In daytime, pathfinders mark touchdown points so the pilot can find a clear and safe landing area.

(3) ***Whiteouts***. If the tactical situation permits, the GTA radio operator advises the pilot of the surface conditions so the pilot can plan how to approach. Using the echelon left or right landing formation reduces the effect of snow clouds (also called "whiteouts") on subsequent landings. The pathfinder plans to stagger aircraft arrivals to let the snow clouds settle.

(4) ***Multiple Landing Sites***. Depending on the mission requirements, climatic conditions, and the expected times of the landings, the pathfinder leader may decide to use multiple landing sites.

(5) ***Night Approaches***. Aircraft making night approaches to snow sites need a visual reference point on the ground such as runway or tactical landing lights. These lights help the pilot judge the angle of descent and rate of closure. He plans the approach to land short of the touchdown point. This ensures that he does not overshoot the point and have to decelerate rapidly in the snow cloud produced by his own aircraft. Approaching short allows the pilot to maintain airspeed after leveling off, and to keep the aircraft in front of the snow cloud until touchdown.

(6) ***Adjustment of Inverted "Y."*** If he coordinates with the flight commander before the landing, the pathfinder can adjust his inverted "Y" forward 10 meters in front of his designated Number 1 touchdown point. This way, the landing site party takes advantage of all usable areas on the site.

d. **Signalman**. The pathfinder leader positions any extra personnel to act as signalmen for aircraft approaching other touchdown points. While aircraft approach and land, he makes sure that signalmen remain in safe areas. Other signalmen should also control the loading of personnel on the aircraft, as instructed by the crew chief or the crew. The technique for landing on snow with a sling load resembles other types of approach, but the pilot hovers at a higher altitude because of the load (A, Figure 4-18, page 4-26). He has a hard time judging the height of the sling load (the height above the ground) as it nears the snow surface. He relies on a signalman to keep him informed. To avoid building up a snow cloud, the pilot puts the load on the ground as fast as he can (B, Figure 4-18, page 4-26).

e. **Sling Load Operations**. The CH-47 requires a sling length of at least 60 feet. Other aircraft allow a shorter sling. Normally an aircraft hovers during hookup and liftoff with a sling load. Doing this above snow produces a snow cloud (B, Figure 4-18, page 4-26). The pilot must expect this and plan for it. In fact, when operating over snow-covered terrain, the pilot can use the most common technique-hovering the helicopter over the load while the ground crew attaches the sling to the hook-or not.

(1) The pilot can land to the left of the load, but close enough for hookup personnel to attach the sling to it.

(2) When ready for liftoff, the pilot initiates a slow, vertical ascent with enough lateral movement to position the aircraft over the load (C, Figure 4-18, page 4-26). He continues a vertical ascent until the load has cleared the ground and he has conducted a hover-power check. When the load clears the ground, the pilot begins accelerating, and continues to climb.

WARNING

Rotor wash increases the risk of frostbite. Make sure you and anyone else on the ground dresses for the conditions and keeps or uses a face mask and goggles.

f. **Static Electricity**. During cold weather, static electricity creates serious problems. Moving an aircraft through the air, brushing snow and ice from an aircraft, or dragging steel cables over the snow can generate static electricity. During external load operations, aviators key the FM radio just before load pickup. This discharges the aircraft's static electrical charge. Because the charge rapidly builds up again, hookup personnel use a grounding device to avoid electrical shock (Chapter 6 provides additional information about the static probe).

g. **Safety**. Accumulated ice on aircraft structural and moving parts presents a danger to nearby ground personnel. The aircraft can accumulate ice up to three-quarters of an inch thick during flight in temperatures and altitudes where icing conditions exist. During flight at less extreme temperatures, this ice begins to loosen and fall off. Ice may shed while the helicopter loses altitude during the landing approach and during touchdown, and pieces of ice shed by the main rotor can fly outward as much as 300 feet. Ground personnel should stay a safe distance away from helicopters during landing and shutdown (after flight in icing conditions), and passengers should not exit until the rotor blades have stopped.

4-19. JUNGLE

Jungle areas impede military operations. Jungle areas promise heat, humidity, rainy seasons, and other weather conditions that reduce aircraft performance.

a. **Communications**. Jungle tends to obstruct military lines of communication. Thick vegetation, irregular terrain, and adverse atmospheric conditions screen radio transmissions. The ground or supported might have to use radio relays. They might also have to staff and mark the CCP. If communications are limited in range, pathfinders might also have to provide GTA communications to advise and direct the pilot to the landing site.

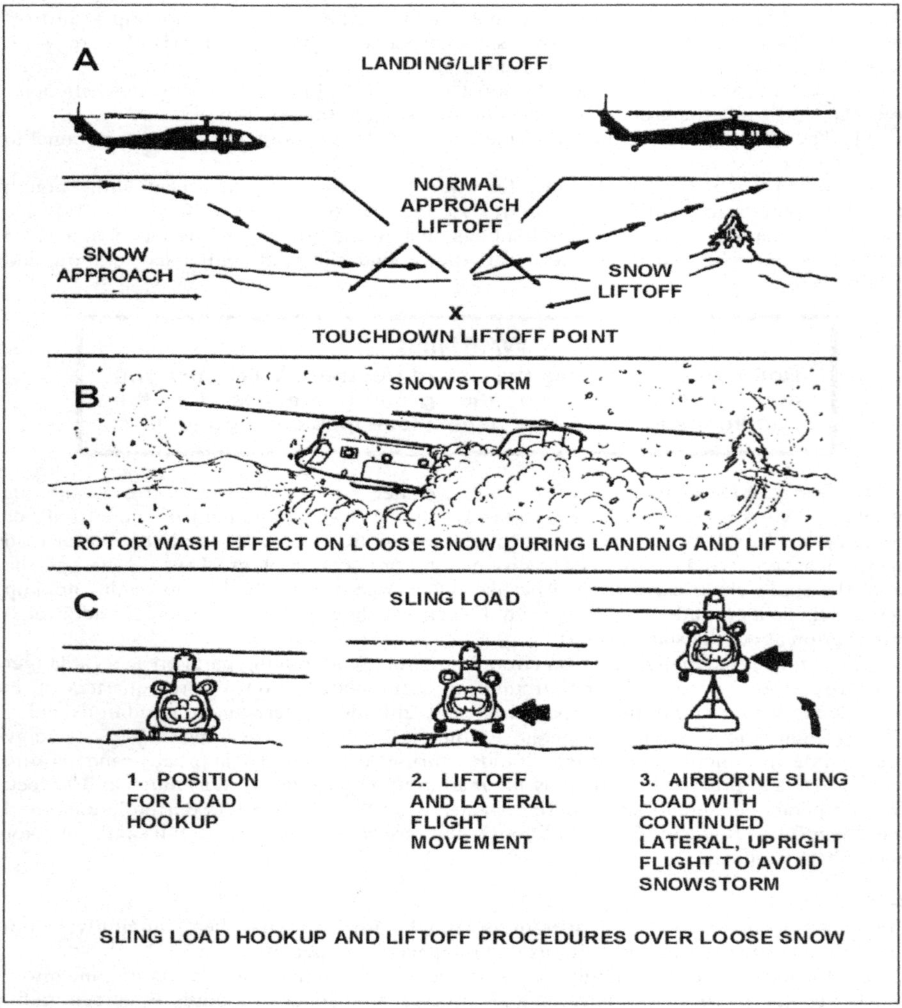

Figure 4-18. Lessening the effects of loose snow on the ground.

b. **Landing Sites**. Jungle conditions mean small landing sites that can handle only a few aircraft at a time. Small landing sites also mean a reduced allowable cargo load (ACL). Pathfinders evaluate surface conditions at the landing site to make sure the aircraft will not sink or bog down in the soil. Then, they survey the site for vines, trees, and other obstructions in the approach path and near the touchdown point.

c. **Navigation**. On an approach to a jungle landing site, the pilot avoids using a high rate of descent. He uses a steep enough angle of descent to just clear any obstacles. He normally uses a ten-to-one obstacle ratio, but for a jungle operation, he should reduce this ratio to no less than five to one. Due to density altitude problems in tropical areas, the aircraft may not be able to develop enough lift to clear tall obstacles. So, the pathfinder leader considers obstacle height on the approach and departure ends. When site size and terrain conditions permit, the pilot might run the liftoffs and landings. However, the small size of a jungle site, soft terrain, or obstacles can keep him from doing so.

d. **Lights**. The tactical situation may restrict the use of lights in nighttime jungle LZ operations.

e. **Security**. Success of the ground unit commander's mission relies on site security. Because jungle terrain provides cover and concealment, landing site security presents a constant challenge. The pathfinder team leader coordinates with the flight commander to set a specific time to light the site.

f. **Liftoffs and Landings**. The pathfinder orients the site to the direction of the wind. He keeps departure obstacle ratios low due to climatic conditions, jungle vegetation, and helicopter's reduced lift capability. Since ground effects reduce the aircraft's lift efficiency, the pilot hovers as low as possible and lingers no longer than necessary.

4-20. DESERT

The typical desert is a dry, barren region, generally treeless and sandy. It suffers environmental extremes, with violent and unpredictable weather changes. Its terrain conforms to no particular model. Frequent clear days offer unequaled visibility and flight conditions, but a sudden sandstorm immediately halts all operations. Successful desert operations require special training, acclimatization, and great self-discipline.

a. **Communications**. In desert operations, the radio offers the best way to communicate. The low, rolling terrain allows good radio range. Due to the increased distances involved in military desert operations, FM radio communications may prove inadequate, especially in the higher FM frequencies. Pathfinders, aircraft, and ground crew must all have high-frequency radio equipment. Sand or dust in equipment or a poor electrical ground cause most communication problems. Due to the increased distances between land force units engaged in desert operations, helicopters may provide air or ground relay or help deploy ground radio rebroadcast facilities.

b. **Navigation**. Many of the conditions experienced in cold weather operations resemble those in desert operations. Pathfinders and pilots find distances and altitudes hard to judge in the desert. The lack of definable terrain features makes navigation difficult, especially at night and over long distances. Also, the sameness of the terrain can influence a pilot to pay less attention to his surroundings. Pathfinders may have to mark and man release points.

c. **Landing Sites**. The climatic conditions in the desert profoundly affect the setup and operation of landing sites. Most importantly, the pathfinder must consider density altitude, wind, and sand (dust). Sand on a landing site can produce brownout conditions similar to those in snowy areas, so the same precautions apply. This makes a rocky area a better landing site than a sandy hollow, depression, or valley.

d. **Wind**. Desert winds generally calm down for an hour or two around sundown. Another calm occurs before sunrise. Other than those times, desert winds can drive dense clouds of dust and sand with hurricane force. Strong winds naturally raise dense clouds of dust and sand, and rapid temperature changes often follow strong winds. The pathfinder leader must consider what times of day the wind will allow him to operate the landing site.

(1) The extreme heat often experienced in the desert also affects the aircraft's ACL. When supporting a ground unit, the pathfinder leader coordinates with the aviation element to determine the ACL for each type of aircraft. Both the minimum distance between aircraft and the size of the landing point increase in desert operations: 100 meters between aircraft, 100-meter-diameter landing points. In daylight hours, ground crew members mark the touchdown points. They paint sandbags a bright color or mark them using some other quick method. Ideally, they use signalmen.

(2) When establishing a landing site, the pathfinder leader considers taxi procedures. When an aircraft must taxi, the pilot moves it into a vertical position as quickly as possible to reduce the amount of sand (dust) the engine sucks in as well as to avoid a brownout. Pilots should avoid taxiing over the same area repeatedly.

e. **Liftoffs**. Pilots will not try a normal liftoff in a sandstorm. Helicopters with wheels and airplanes should make a running-type takeoff. Helicopters with skids should make a maximum performance liftoff.

f. **Landings**. When they can, pilots should use a running-type landing to reduce sand intake. If a pilot can make a running landing, he keeps the touchdown roll to a minimum to keep from overloading the landing gear. If the terrain does not permit a running landing, the pilot lands at a greater than normal angle. He should never land from a hover.

g. **Safety**. Ground crew personnel should wear clothing that will protect them against the sand blown around by the rotor wash. Each person on the ground should take special care to keep the sand out of his eyes, ears, nose, and mouth. Goggles, earplugs, and cloth masks provide adequate protection for facial areas. Other ground crew procedures resemble those for cold weather operations.

4-21. MOUNTAINS

Mountains have rugged, divided terrain with steep slopes and few natural or man-made lines of communication. Weather fluctuates seasonally from extreme cold, with ice and snow, to extreme heat. Also, it can switch between the two extremes very quickly. This unpredictability greatly affects operations.

a. **Communications**. Mountainous terrain often limits or restricts communications. To maintain communications within the AO, aircraft may have to limit operations to the vicinity of the unit. Other aircraft can serve as radio relay stations. Pathfinder units may also have to set up radio relays at the RP, CCP, or both.

(1) Mountain conditions challenge aviators in pathfinder operations more than any other conditions. For precise flying in mountainous areas, pilots need large-scale terrain maps.

(2) Since intervening terrain degrades GTA communications, providing navigational aid and control over extended ranges might prove difficult.

b. **Wind**. The main weather hazard in the mountains is wind. Even moderate winds (11 to 20 knots) can produce significant turbulence over mountain ridges. Predicting wind conditions is difficult. The windward side of a mountain maintains a steady direction of airflow, though the strength of the wind may vary. The leeward side has turbulent winds with strong vertical currents. This turbulence might prevent assault landings and require pilots to fly at higher altitudes. This naturally increases the risk of detection and destruction.

c. **Density Altitude**. In the mountains, density altitude can vary a lot between PZs and LZs. It can also vary greatly from one time of day to another. It normally peaks in the late afternoon, and drops to its lowest point at dawn.

d. **Navigation**. In the mountains, the helicopter offers the best way to rapidly move forces. In the offense, air assault operations can insert forces into the enemy's rear area and bypass or envelop his defenses. In the defense, helicopters can move reinforcements and reserves rapidly.

e. **Landing Sites**. Mountainous regions offer few, if any airfields for fixed-wing aircraft, and few LZs suitable for multiple helicopters.

(1) If the enemy situation allows, pathfinders set up LZs on the windward side of the mountain, since that side offers more stable winds.

(2) When they can only find LZs designed for single aircraft, planners increase in-flight spacing. This places an extra load on each crew. When conducting multiship operations into a small LZ, the pathfinder controller should allow sufficient time between liftoff and landing for the turbulent air generated during the departure of the previous helicopter to stabilize. Otherwise, the pilot of the incoming craft will experience that turbulence and lose lift.

(3) A pilot must touch down very carefully on the typical small, rough, sloped mountain LZ. Depending on the angle of the slope and on the aircraft's available torque, the pilot might be able to make a normal slope landing. Pilots of larger craft, such as cargo helicopters, may have trouble positioning the entire fuselage in the available area. Once the cockpit extends over the landing area, the pilot cannot see the ground. He must rely on the crew chief and signalman to direct him.

(4) During a mountain approach to an LZ surrounded by uneven terrain, the pilot has a hard time determining the actual aircraft altitude and rate of closure. Where the terrain slopes up to the LZ, a visual illusion occurs. The pilot may think he is flying too high and closing too slowly. If the terrain slopes down to the LZ, he may feel he is flying too low and closing too fast. Employing a signalman on the ground gives the pilot a visual reference to adjust his controls. He may need more than one signalman.

f. **Site Assessment**. Pathfinders should determine the following information while reconnoitering and selecting a mountain site:

(1) The size, slope, amount of surface debris, and the area covered by shadows and obstacles in and around the site.

(2) The approximate direction, speed, and characteristics of the wind.

(3) The inbound route, if necessary. When the pilot cannot land due to a steep slope, the aircraft may terminate at a hover to off-load troops and supplies.

(4) The departure route. Departure routes should orient into the wind and over the lowest obstacles.

Section VII. APPROACH PATH CONSIDERATIONS

Pilots should try to land their aircraft into the wind; however, the terrain and its effect on the wind may require a crosswind landing. If so, the pilot for single-rotor helicopters should plan the approach so that the wind blows from the left side of the aircraft. This helps the pilot overcome the effects of torque, reduces power requirements, and helps him control the heading. Other considerations include vertical air currents, escape routes, terrain contour and obstacles, and the position of the sun.

4-22. VERTICAL AIR CURRENTS

Updrafts on the approach path make landing easier. However, severe vertical air currents (updrafts or downdrafts) may require the pilot to approach downwind.

4-23. ESCAPE ROUTES

The pathfinders and pilots should plan one or more escape routes along the approach path for the pilot to use if he must go around and try the approach again.

4-24. TERRAIN CONTOUR AND OBSTACLES

The height of terrain and obstacles along the approach path should permit the pilot to conduct a shallow approach angle into the landing site. When possible, the pathfinders select a landing point on or near the highest terrain feature.

4-25. POSITION OF THE SUN

Though the pilot first considers wind direction and nature of the terrain when choosing the approach path, he must also consider the location of the sun and shade relative to the approach path. To keep the pilot from having to adjust from one light condition to another, the pathfinder makes sure that if the landing point falls in a shaded spot, that the whole approach path also does. When the sun rises or falls to just above the horizon, avoid using an approach path that faces directly into it.

CHAPTER 5
EXTERNAL LOADS

Carrying cargo and equipment outside the helicopter eliminates many of the obstacles that other modes of transportation suffer. Helicopters move cargo by external sling load when--

- *The cargo compartment cannot hold the load.*
- *The load exceeds the helicopter's internal load limitation.*
- *The ground crew must load or unload the cargo at once.*
- *Landing zone conditions prevent the aircraft from touching down.*

Pathfinder-qualified soldiers prepare to organize and control external load pickup or drop-off sites as an integral part of LZ operations. The supported unit provides a detailed load plan, to include rough weights and sequences of load movement. This ensures the correct and rapid movement and placement of cargo.

5-1. LANDING POINTS

All sling-loading operations use Size 4 or 5 landing points (80- or 100-meter diameter) regardless of the type or size of the helicopter. Area conditions, such as a dusty surface or obstacles, often require pathfinders to increase the minimum spacing between loads. This reduces the number of helicopters that can safely operate at the site at the same time. It also reduces the overall speed of the operation.

5-2. TYPES OF LOADS

All external loads fall into three types: high density, low density, and aerodynamic. Each exhibits unique characteristics in flight. Pathfinders determine the category, size, and weight of the load during the planning phase of the operation.

a. The high-density load offers the best stability

b. The low-density load offers the least stability.

c. The aerodynamic load lacks stability until the airstream stabilizes the load.

5-3. UNIT RESPONSIBILITIES

Most sling-load operations involve three units. Each has pathfinders who perform specific functions.

a. **Supported Unit**. The supported unit moves equipment and rigs the loads. Pathfinders in the supported unit check the weight, rigging, and positioning of all external loads to ensure helicopter safety. Ideally, the supported unit provides hookup personnel for individual loads. In the supported unit, pathfinders--

(1) Coordinate in advance with the supporting unit.

(2) Rig the loads.

(3) Furnish slings, straps, clevises, and any other sling-load equipment required for the move.

(4) Check for improper rigging and weight in excess of the aircraft's allowable cargo load.

b. **Supporting Unit**. This is the aviation unit that will fly the loads.

(1) Provide advice and technical help to the supported unit, as required.

(2) Ensure that the loads fall within the transporting aircraft's ACL.

c. **Pathfinder Unit**. In the pathfinder unit, pathfinders--

(1) Provide advice and aid to both the supporting and supported units.

(2) Provide expertise in the planning and execution of both pickup zone (PZ) and helicopter landing zone (HLZ) operations.

(3) Supervise the rigging and inspection of all the loads.

(4) Provide ground guidance and air traffic control during the slingload.

(5) Ensure that the loads fall under the transporting aircraft's ACL.

d. **Allowable Cargo Load**. The ACL depends on the type of aircraft, the age of the airframe, the altitude above sea level, the temperature, the humidity, and the aviation unit's SOP.

5-4. EQUIPMENT

Cargo nets and slings make up an essential part of the external load operation. During an inspection, they require the same level of attention that the cargo receives. Any evidence of frayed or cut webbing justifies replacement of the affected component. When they assemble slings, pathfinders should avoid sewing up torn slings or substituting nonstandard parts in the field. Slings must meet the critical strength requirements specified in FM 10-450-3. The Army's inventory includes a variety of equipment adapted or designated for use in sling-load operations.

a. **Aerial-Delivery Slings**. ADSs were originally designed to deliver heavy loads by air (Table 5-1). They have been adapted for use in air assault operations. ADSs come in a variety of sizes and strengths, as shown in FM 10-450-3, Chapter 6, Table 8-2, and Appendix B.

b. **Hitches**. When connecting ADS to metal air items or directly to the load, loaders use one of the following hitches (Figure 5-1):

(1) ***Choker Hitch***. Pull the free-running end of the sling around the point of attachment. Draw it between the loops of the sling's standing end. After making sure that the cotton buffer is in its proper place, "milk" down the keeper on the standing end to secure the sling.

(2) ***Basket Hitch***. Separate the loops of the sling at one end. Place the sling over the suspension point. Ensure that the cotton buffer is in its proper place. To secure the sling, "milk" down the keeper towards the suspension point.

USAGE	LOOP SLINGS	VERTICAL PENDANT POUNDS	THICKNESSES	AVAILABLE LENGTHS (IN FEET)	
Pendant	2-Loop [1,2]	8,900	4	3 Feet	16 Feet
				9 Feet	20 Feet
				11 Feet 12 Feet	120 Feet
	3-Loop	13,500	6	60 Feet	140 Feet
	4-Loop	17,800	8	3 Feet	16 Feet
				9 Feet	20 Feet
				11 Feet	28 Feet
				12 Feet	
	6-Loop	27,000	12	60 feet	120 feet
	9-Loop	42,000			
Part of sling set	2-Loop [1,2]	5,600	4	3 Feet	16 Feet
				9 Feet	20 Feet
				11 Feet	120 Feet
	3-Loop	8,500	6	60 Feet	140 Feet
	4-Loop	11,200	8	3 Feet	16 Feet
				9 Feet	20 Feet
				11 Feet	28 Feet
				12 Feet	
	6-Loop	17,000	12	60 Feet	120 Feet

1 Identified by colored thread stitched lengthwise down the middle of the strap.

2 Three-foot donut ring tensile strength for this sling equals 10,000 pounds. Using dual rings increases tensile strength to 17,500 pounds.

Table 5-1. Aerial delivery specifications for the Type XXVI sling.

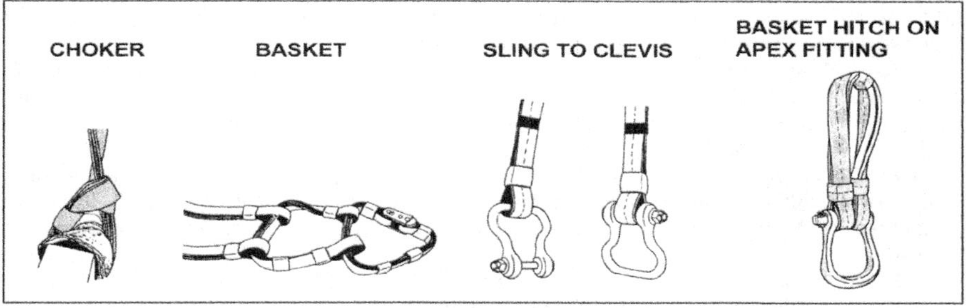

Figure 5-1. Hitches.

c. **Nets and Containers**. The Army has many types of cargo containers. However, the 5,000- and 10,000-pound cargo nets and the A-22 cargo bag are the ones most often used to transport cargo externally. Chapter 7, FM 10-450-3 describes how to inspect both the large cargo net and the A-22 cargo bag (Figure 5-2). It also provides rigging instructions. Avoid overloading the nets; use them with loads that fall within the aviation unit's prescribed limits. Pick up the nets rather than dragging them across the ground, since dragging them can cause them to snag on something and damage the net or the thing it snags. Use a canvas insert when carrying items small enough to slip through the netting:

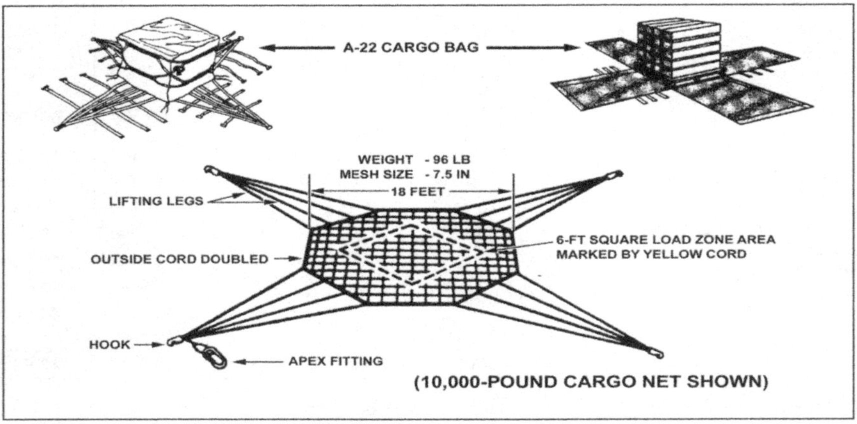

Figure 5-2. Cargo nets and bag.

(1) ***Small (5,000-Pound Capacity) Cargo Net***. This olive drab net can carry up to 5,000 pounds or 125 cubic feet of cargo (NSN 1670-01-058-3811, LIN NO2776).

(2) ***Large (10,000-Pound Capacity) Cargo Net***. This larger net can carry up to 10,000 pounds or 380 cubic feet of cargo (NSN 1670-01-058-3810, LIN NO 2708). It is 18 feet wide and can transport boxed or bulky loads.

(3) ***A-22 Cargo Bag***. The A-22 cargo bag, with or without its canvas cover, can externally transport standard palletized loads, loose cargo, ammunition, oil drums, and other general items whose total weight falls under 2,200 pounds (Figure 5-3).

(4) ***Suspension Clevises***. Clevises come in three sizes.

(a) *Large Suspension Clevis*. This clevis (NSN 4030-00-090-5354) has a rated capacity of 12,500 pounds (pendant) with a 7,875-pound sling-to-lifting provision point of attachment. Adding more large clevises as attaching points increases rated capacity as follows:

- Two large clevises increase rated capacity to 15,750-pounds.
- Three large clevises increase rated capacity to 23,625 pounds.
- Four large clevises increase rated capacity to 31,500 pounds.

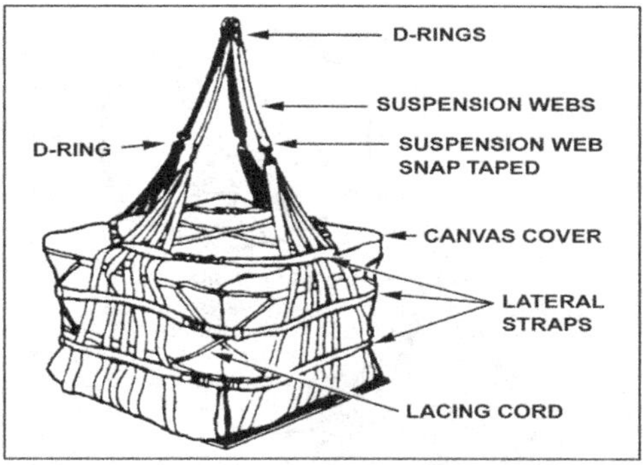

Figure 5-3. A-22 cargo bag.

(b) *Medium Suspension Clevis*. This clevis (NSN 1670 4030-00-678-8562, Figure 5-4, page 5-6) has a rated capacity of 6,250 pounds (pendant) with a 3,750-pound sling-to-lifting provision. Adding more medium clevises as attaching points increases rated capacity as follows:

- Two medium clevises increase rated capacity to 7,500 pounds.
- Three medium clevises increase rated capacity to 11,250 pounds.
- Four medium clevises increase rated capacity to 15,000 pounds.

(c) *Small Suspension Clevis*. This clevis (NSN 1670 4030-00-360-0304) has a rated capacity of 6,250 pounds (pendant) with a 3,750-pound sling-to-lifting provision. Adding more small clevises as attaching points increases rated capacity as follows:

- Two small clevises increase rated capacity to 7,500 pounds.
- Three small clevises increase rated capacity to 11,250 pounds.
- Four small clevises increase rated capacity to 15,000 pounds.

d. **Reach Pendants**. A reach pendant is a synthetic rope assembly with an attached, stiffened reach tube and a loop on each end. The built-in reach tube enables the hookup man to place the pendant's top eye on the helicopter cargo hook while the helicopter hovers at a higher distance over the load (Figure 5-5, page 5-6). Two reach pendants are authorized for use with slingloads. To use either pendant with a sling set, remove the sling set apex fitting pin; place the pendant's lower eye in the apex fitting; and reinstall the apex fitting pin.

(1) ***11,000-Pound-Capacity Reach Pendant***. The 11K, NSN 4020-01-365-3115, part number DSG-5-11K, measures about 5 feet long and has an 11,000-pound safe working load capacity. The top eye is black with a smaller diameter loop, while the bottom eye is green with a larger diameter loop. The safe working load capacity is stamped on the reach tube.

(2) ***25,000-Pound-Capacity Reach Pendant***. The 25K, NSN 4020-01-337-3185, part number BOS-14-K7, measures about 5 feet long and has a 25,000-pound safe working load capacity. The top eye is black with a smaller diameter loop, while the bottom eye is also black, but has a larger diameter loop. The safe working load capacity is stamped on the reach tube.

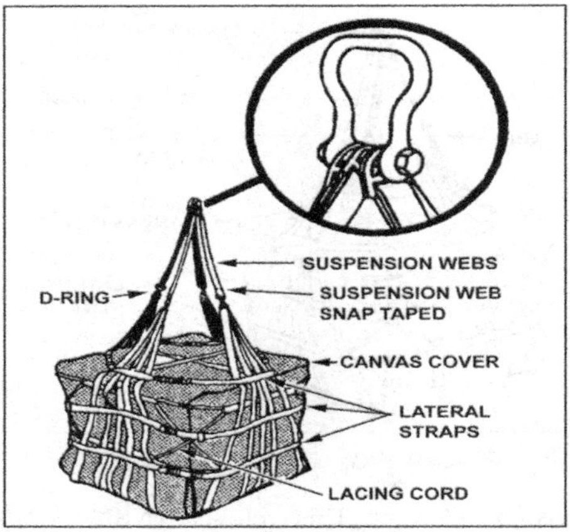

Figure 5-4. Upper sling and medium clevis.

CAUTION

The older, 25,000-pound capacity reach pendant, issued with an aircraft aerial recovery kit (ARK), P/N 1670EG093-1, is no longer authorized for use in slingload operations.

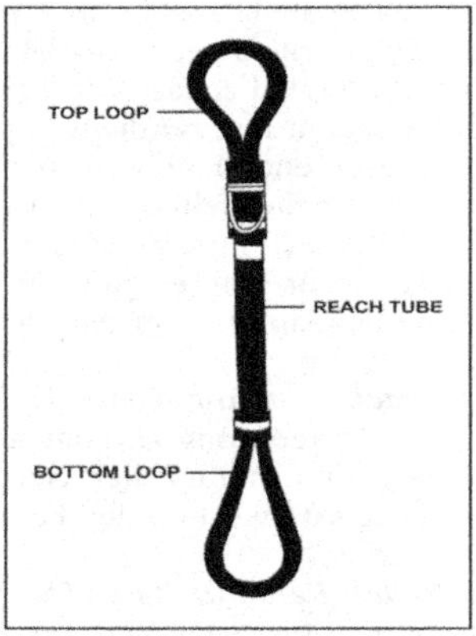

Figure 5-5. Reach pendant.

(3) ***Inspection***. Inspect the reach pendants *before and after* use. Check for cuts and tears in the nylon-urethane plastic sheath on each loop. If the white strength member (third layer) shows, remove the pendant from service.

(4) ***Cleaning***. Clean the reach pendant with a mixture of warm water and mild dish or laundry detergent. You can use mineral spirits to remove oil and grease. You can treat the top and bottom eyes with silicone spray. However, do not use silicone spray on the reach tube.

(5) ***Storage***. Store the reach pendants in a clean, dry area out of direct sunlight. Prolonged exposure to sunlight will deteriorate the strength of reach pendants.

CAUTION

Avoid getting silicone spray on the reach tube.

Avoid using chemical cleaners on reach pendants. Chemicals may weaken the strength members of the pendant. If a pendant becomes contaminated with chemicals, remove it from service.

e. **Polyester Roundslings**. Use polyester roundslings as the primary vertical pendant (Figure 5-6, page 5-8). You can use the three hitches described in Chapter 6 to attach roundslings to the load. The lifting capacity of polyester roundslings varies with the size of the sling and the type of hitch used to attach the load. Each sling has two identification tags permanently sewn to the eye and eye sleeve. These identify the size and capacity of the roundsling as well as other information needed for its safe use. Roundslings are also color-coded by size. Table 5-2 (page 5-8) lists roundsling lengths and lift capacities.

(1) ***Storage***. Store roundslings in a clean, dry, cool area out of direct sunlight. Prolonged exposure to sunlight will deteriorate the strength of roundslings.

(2) ***Inspection***. Inspect each polyester roundsling before and after every use. Remove it from service if you find any of the following:

(a) Missing or unreadable identification tags.

(b) Acid or alkali burns.

(c) Melted, charred, or weld-splattered portions.

(d) Any holes, tears, cuts, snags, embedded particles, broken or worn stitching, or abrasive wear that exposes the core fibers.

(e) Knots in any part of the roundsling.

(f) Distortion, excessive pitting, corrosion, or broken fitting(s).

(g) Any other condition that causes doubt as to the strength of the roundsling.

CAUTION

Avoid dragging roundslings on the floor or over rough surfaces. Never twist them or join them together with knots.

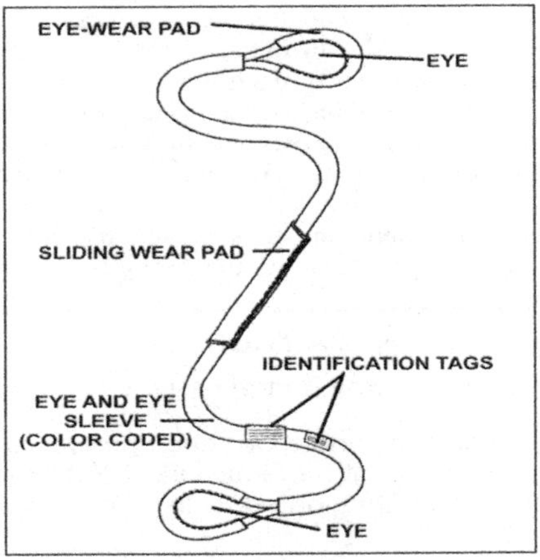

Figure 5-6. Polyester roundslings.

PART NO.	LENGTH (IN FEET)	COLOR	LIFT CAPACITY			WEIGHT (IN POUNDS)
			CHOKE	VERTICAL	BASKET	
PRS2E008	8	GREEN	4,200	5,300	10,600	4
PRS2E017	17	GREEN	4,200	5,300	10,600	10
PRS3E008	8	YELLOW	6,700	8,400	16,800	5
PRS3E017	17	YELLOW	6,700	8,400	16,800	11
PRS5E030	30	RED	10,600	13,200	26,400	26
PRS7E065	65	BLUE	17,000	21,200	42,400	75
PRS7E070	70	BLUE	17,000	21,200	42,400	81

Table 5-2. Safe working loads (lift capacities) of polyester roundslings.

5-5. SERVICE LIFE OF AERIAL-DELIVERY SLINGS

The first person to use a sling must date-stamp it with the calendar or Julian date in 1-inch letters. He can use orange-yellow parachute-marking ink, strata blue parachute-marking ink, or an orange-yellow tube-type marker. He marks near the first keeper at both ends of the sling. This date determines the date of the next inspection. Every six months, the current user reinspects the sling, strikes through the last date in the same color it was written in and, in either of the other two colors, he marks the date he reinspected the sling. For example, if the original user marked the old date in strata blue, the next inspector marks through that date, also using strata blue ink or marker. Then, he uses stencils the current date in orange-yellow, or vice versa. *Every single user* inspects *every single sling* before and after *every single use*. If the condition of the sling seems questionable, he removes the sling from service (Figure 5-7).

a. **General Inspection**. When inspecting nylon air items, look for any sign of the following. Note that if you find more than three consecutive broken or loose stitches, or five or more broken or loose stitches overall in the sewn portion, the item is unserviceable. After rigging the load with any nylon air item, put cotton buffers in place to prevent any nylon-to-nylon or nylon-to-metal contact:

- Inspection date that has already passed (an inspection is overdue).
- Foreign matter or chemicals such as mildew, paint, or grease.
- Cuts.
- Frays.
- Burns.
- Broken stitches.
- Missing cotton buffers, sliding keepers, or permanent keepers.
- Rust.

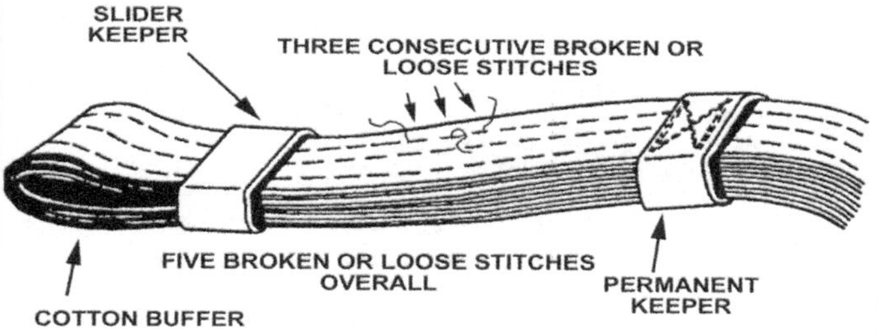

Figure 5-7. Unserviceable slings.

b. **Cargo Strap**. The A7A cotton or nylon cargo strap measures 188 inches long and has a rated capacity of 500 pounds. A friction adapter located on one end of the strap has a thick-lipped metal floating bar. Supply issues this strap with one metal D-ring. Inspect this piece of equipment for cuts or frays.

c. **Cargo Tie-Down Equipment**. Check the tie-downs for serviceability.

(1) ***CGU-1B Tie-Down Strap***. The CGU-1B cargo tie-down device has a rated capacity of 5,000 pounds. You can adjust the length of this device.

(2) ***15-Foot Tie-Down Strap***. The 15-foot cargo tie-down strap, issued with a quick-fit strap fastener, has a rated capacity of 5,000 pounds.

(3) ***Load Binders***. The two load binder types are rated for 10,000 pounds and 5,000 pounds. The 10,000-pound capacity load binder has its rating stamped on the side.

d. **Metal Air Items**. Sling-load operations commonly require the use of the following metal air items. Thoroughly inspect them for rust, stripped threads on the nuts or bolts, burrs, cracks, bent or twisted metal, or oil. When using any clevis assembly, tighten the nut hand-tight only. Then, tape both ends of the nut and bolt assembly to prevent slippage during use. Use only case-hardened nuts and bolts--never mix items.

The bolt heads have case-hardened marks such as ticks, numbers, letters, or a combination of all three:

(1) ***Inspection***. Use the Type IV link assembly (NSN 1670-00-783-5988) to build a 3-foot donut or to connect one ADS to another (Figure 5-8). This link assembly has a rated capacity of 12,500 pounds. When inspecting the Type IV link assembly, look for the following deficiencies:

- Hard to rotate or irregularly rotating aluminum buffers.
- Bent or cracked posts.
- Bent slide connectors.
- The absence of a metallic "click" when it locks.

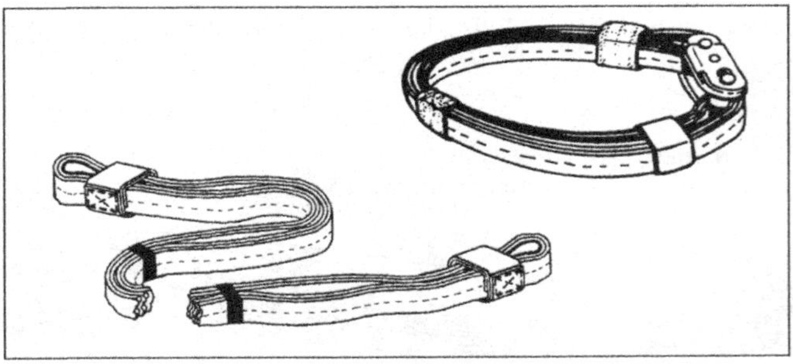

Figure 5-8. Three-foot apex (donut) ring.

(2) ***Points of Attachment***. In sling-load operations, the clevis assemblies serve as points of attachment from the aircraft to the load.

(3) ***Tightening.*** When using any clevis assembly, tighten the nut by hand only. Tape both ends of the nut and bolt to prevent slippage during use. Choose only case-hardened nuts and bolts. Never mix items. Case-hardened marks include ticks, numbers, and letters. They will be located on the bolt head.

e. **Large-Capacity Sling Sets**. The new 10,000-pound and 25,000-pound capacity sling sets are similar, except for a few minor differences. All components have identifying marks. You may only exchange apex fittings between sets. Take care not to mix up the other components. Table 5-3 compares these two large-capacity sling sets. Figure 5-9 shows one.

	TYPE SLING:	10,000-POUND SLING	25,000-POUND SLING
	CAPACITY:	10,000 POUNDS	25,000 POUNDS
APEX FITTING	COLOR:	BRUSHED ALUMINUM	GOLD STEEL
	PIN DIAMETER:	1 1/8-INCH DIAMETER	1 1/2-INCH DIAMETER
	WEIGHT:	4 1/2 POUNDS	10 POUNDS
SLING ROPE	COLOR:	OLIVE DRAB	BLACK
	LENGTH:	12 FEET	12 FEET
	DIAMETER:	7/8 INCH	1 1/4 INCH
	CHAIN LINKS-QUANTITY:	110 to 115 LINKS	86 TO 88 LINKS
	NSN:	1670-01-027-2902	1670-01-027-2900
	TOTAL WEIGHT:	52 POUNDS	114 POUNDS

Table 5-3. Large-capacity sling sets.

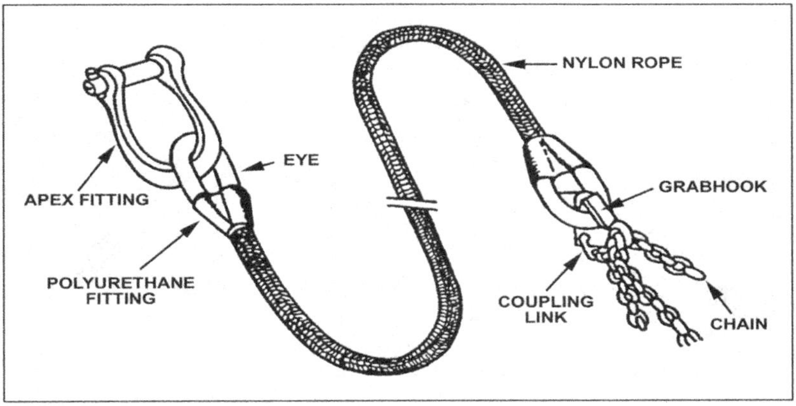

Figure 5-9. 25,000-pound capacity sling set.

CAUTION

Each sling set has four legs. Each leg has a rated capacity of one-quarter of the total capacity of the set. On some loads, you will use up to six legs. However, remember that adding two legs does not increase the rated capacity of the entire set.

(1) The nylon rope assembly for each set has an interwoven eye located at each end.

A polyurethane fitting covers the eye to protect the leg from abrasion and ultraviolet radiation (Figure 5-9). Each double-braided rope connects to a grabhook assembly. Figure 5-10, Figure 5-11, and Figure 5-12 show a coupling link, sling leg-numbering sequence, and a grabhook, respectively. Though the grabhooks for the two sets look alike, you cannot interchange them, because they have different ratings.

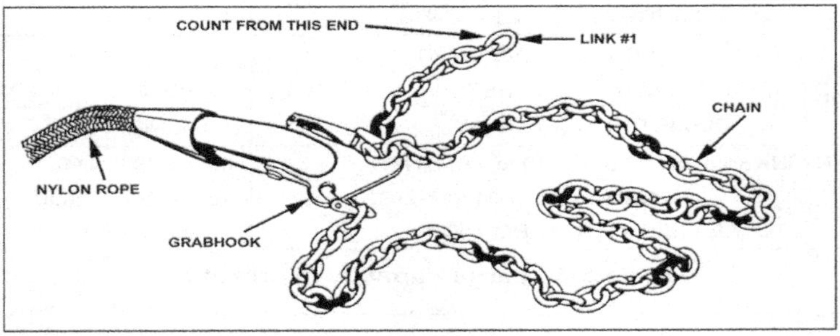

Figure 5-10. Coupling link.

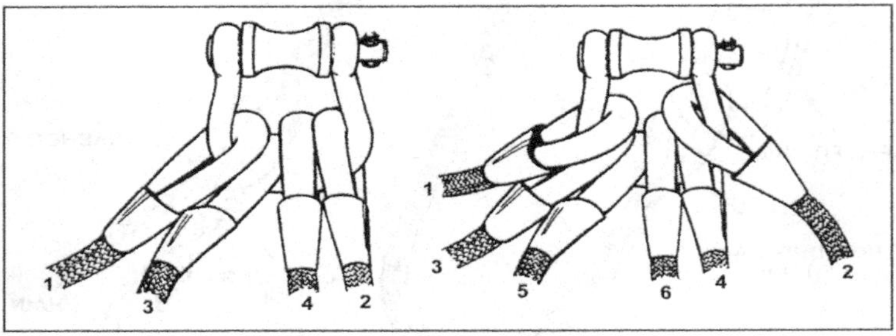

Figure 5-11. Sling leg-numbering sequence.

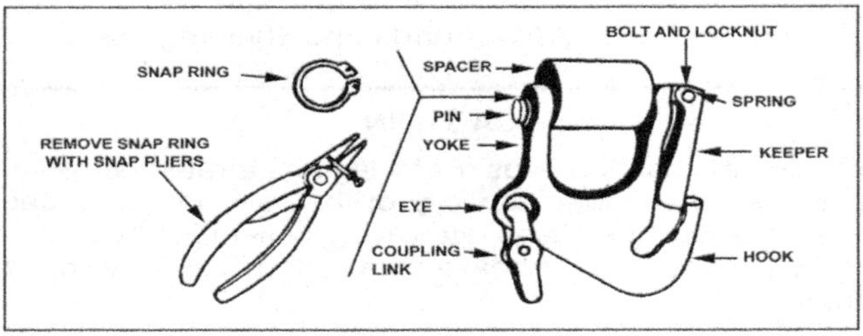

Figure 5-12. Grabhook.

(2) Chapter 6, FM 10-450-3 discusses how to inspect the rope sling sets. Appendix C, FM 10-450-3 provides sling-conversion tables.

(3) Secure the cross pin on each apex fitting with a 3/8-inch bolt, a castellated nut, and a cotter pin. Secure the dome-shaped nut to the apex wrench-tight.

5-6. AIRCRAFT LOAD LIMITATIONS

The structural strength of the cargo hook assembly determines the maximum weight that any aircraft can carry with an external slingload. In most cases, the tensile strength of the hook does not limit the weight that an aircraft can lift--the allowable cargo load does. In fact, the capacity of the cargo hook assembly usually exceeds the ACL.

a. **Tensile Strength**. Cargo hook tensile strengths for US Army aircraft show--

(1) UH-1H/UH-1N Iroquois-4,000 or 5,000 pounds. See Figure 5-13.

(2) UH-60A/60L Black Hawks-8,000 or 9,000 pounds. See Figure 5-14, page 5-14.

(3) CH-47C/D Chinooks-20,000 or 26,000 pounds. See Figure 5-15, page 5-14.

b. **Capacities**. CH-47D's fore and aft cargo hooks can each lift 17,000 pounds; together, they can lift 25,000 pounds. The CH-47C's center cargo hook can lift 20,000 pounds. The CH-47D's center cargo hook can lift 26,000 pounds.

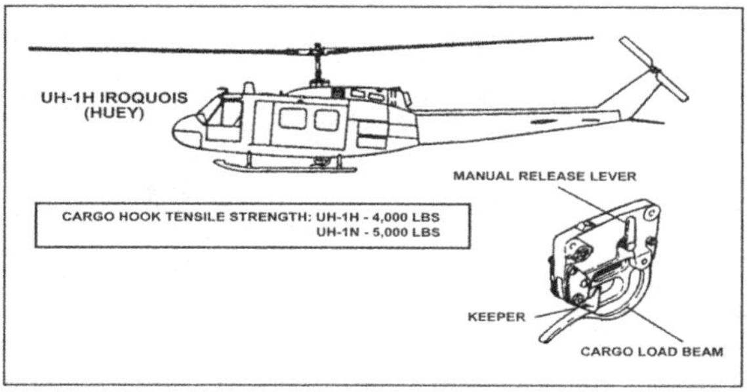

Figure 5-13. UH-1H Iroquois and its cargo hook.

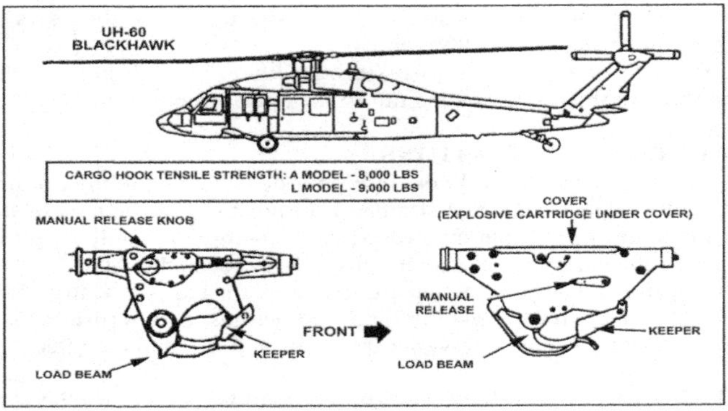

Figure 5-14. UH-60 Black Hawk and its cargo hooks.

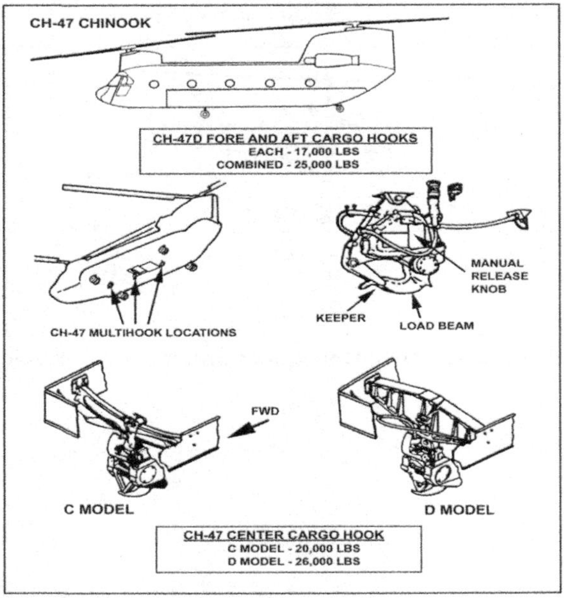

Figure 5-15. CH-47 Chinook and its cargo hooks.

c. **UH-1H**. When transporting external loads with an UH-1H, use a nylon donut or web ring to attach the load to the aircraft. The cargo hook on the UH-1H is stationary;

using an apex with a heavy load would bind the hook and shear it off. The cargo hook tensile strengths for the UH-1H is 4,000 pounds; for the UH-1N, it is 5,000 pounds.

d. **UH-60 Black Hawk**. When using the 10,000-pound apex to secure an external load to the UH-60, you must also use the spacer. When you use the 25,000-pound apex, you must also manually operate the cargo hook. *Never* use a donut or web ring on an UH-60. The web ring will bind on the hook and prevent the crew from releasing the load in an emergency. The tensile strength of the UH-60 cargo hook system is, for the UH-60A, 8,000 pounds; and for the UH-60L, 9,000 pounds.

5-7. STANDARD WEIGHTS

When using an UH-60 Black Hawk for airlift, coordinate closely with the aviation unit for the ACL. For standard POL weights (for external loads only), see Table 5-4 on page 5-16. Standard weights for vehicles and artillery follow:

a. **Vehicles**.

(1) ***M998/M1038 Truck, Cargo, 1 1/4-ton (HMMWV).***
-- 5,200 pounds empty.
-- 7,700 pounds loaded.

(2) ***M966 TOW Missile Carrier (HMMWV)***.
-- 6,050 pounds empty.
-- 8,200 pounds loaded.

(3) ***M416 1/4-Ton Trailer***.
-- 580 pounds.

(4) ***M101A2 3/4-Ton Trailer***.
-- 1,350 pounds.

(5) ***M105A2 1 1/2-Ton Trailer***.
-- 2,750 pounds.

(6) ***M35A2 2 1/2-Ton Truck***.
-- 12,000 pounds (add 500 pounds if equipped with a winch).

(7) ***M149 1 1/4-Ton Water Trailer***.
-- Empty-2,540 pounds.
-- Full-6,060 pounds.

(8) ***M149A1 1 1/4-Ton Water Trailer***.
-- Empty-2,540 pounds.
-- Full-6,060 pounds.

(9) ***M149A2 1 1/4-Ton Water Trailer***.
-- Empty-2,800 pounds.
-- Full-6,320 pounds.

b. **Artillery Equipment**.

(1) ***M101 105-MM Howitzer***.
-- 4,600 pounds (add 300 pounds if equipped with shields).

(2) ***M102 105-MM Howitzer***.
-- 3,160 pounds (add 170 pounds for section equipment).

(3) ***105-MM Ammunition***.
-- 60 pounds (each box).

(4) ***105-MM Ammunition***.
-- 47 pounds (each carton).

FUEL	55-GALLON DRUM	500-GALLON BLIVET
Gasoline (MOGAS)	404 pounds	3,400 pounds
Gasoline (JP4/JP8)	410 pounds	3,500 pounds
Diesel fuel	457 pounds	3,800 pounds
Lube oil (30 weight)	479 pounds	4,000 pounds

Table 5-4. POL for external loads only.

5-8. AIR ITEMS REQUIRED FOR COMMON STANDARD LOADS

Pathfinders require several types of expendable rigging supplies to complete the rigging of the loads discussed in this paragraph. These supplies include 1/4-inch cotton webbing, 1/2-inch diameter rope, Type III nylon cord (550-pound test), pressure-sensitive tape, cellulose wadding or paperboard energy-dissipating material, and canvas or felt padding. Pathfinders should obtain sufficient supplies of these items before rigging the loads. To rig loads with ADSs (aerial delivery slings) and with more than one suspension point, twist the ADS once for each 3 feet of sling length. This reduces vibration in the sling during flight. The nylon and chain multileg sling sets and the 10,000- and 25,000-pound capacity sling sets do not require the twists. FM 10-450-3 discusses preparation and rigging for the following loads in detail:

a. **Net**. A 10,000-pound-capacity nylon cargo net (18 feet by 18 feet) requires one A7A cargo strap.

b. **Perforated Steel Planking**. Perforated steel planking (PSP) requires--

- Two 16-foot, two or three-loop ADS.
- One 3-foot ADS with one Type IV link assembly (for doughnut).

c. **Fuel Drums**. One or two fabric fuel drums (blivets), each of which contains 500 gallons of fuel and a 10,000-pound-capacity sling set.

d. **Concertina Wire**. The items required to move this load depend on the amount of concertina wire in the load.

e. **Cargo Bag**. The A-22 cargo bag consists of--

(1) One A-22 kit.
(2) One A7A cargo strap.
(3) One 8-foot, two- or three-loop ADS.
(4) One 3-foot, two- or three-loop ADS with one Type IV link assembly (for doughnut).

5-9. SLINGLOAD THEORY

The behavior of an external load in flight can greatly affect the performance of the aircraft carrying it. High drag coefficients reduce airspeed. Therefore, whoever prepares the load must try to reduce the drag of the load on the aircraft. This means that the task takes longer or does not get finished if the allotted time for the task expires. A high drag coefficient can also endanger the aircraft and crew. Because of this, the pilot must "punch" if he thinks that continuing to fly the load could endanger his crew or aircraft. To stabilize a load, the loaders should--

a. **Addition of Weight to the Load**. The heavier a load, the less air pressure will disturb it. Thus, carrying heavier loads assures greater stability. However, make sure the load does not exceed the rated capacity of the equipment or the ACL of the aircraft.

b. **Streamlining of the Load**. Long, symmetrical loads fly crosswise to the direction of flight. This causes a lot of drag on the aircraft. Loads tend to stabilize if the CG (center of gravity) is located in the first one-third of the load. Either adjust the load or, if needed, add weight to move the CG toward one end or the other. The heavier end of the load will "seek" the direction of flight and the load will stabilize. The lighter tail end of the load will act just like the fins on a dart.

c. **Airspeed of the Aircraft**. The least desirable method is to have the aircraft fly slow to try to keep the load from destabilizing. This burns extra fuel and takes more time to do less work. Prepare the loads so that the aircraft can fly safely at speeds of 60 knots or more.

d. **Sling Length**. Lengthening the slings that attach the load to the aircraft reduces the load's stability in flight. The shorter, the better, as long as the sling measures at least 6 feet long. Also, the more vertical the attached sling, the less stress on those that are more horizontal. Figure 5-16 (page 5-18) shows how sling angle affects load stress.

e. **Effect of Vertical Sling**. A vertical sling carrying 3,000 pounds has 3,000 pounds of stress on it. That means the stress equals the weight of the load.

f. **Effect of 45-Degree Sling**. A 45-degree sling carrying 3,000 pounds has 4,242 pounds of stress on it. That means the stress equals nearly one-and-a-half times the weight of the load.

g. **Effect of 5-Degree Sling**. A 5-degree (almost horizontal) sling carrying 3,000 pounds has 34,419 pounds of stress on it. That means the stress equals more than ten times the weight of the load.

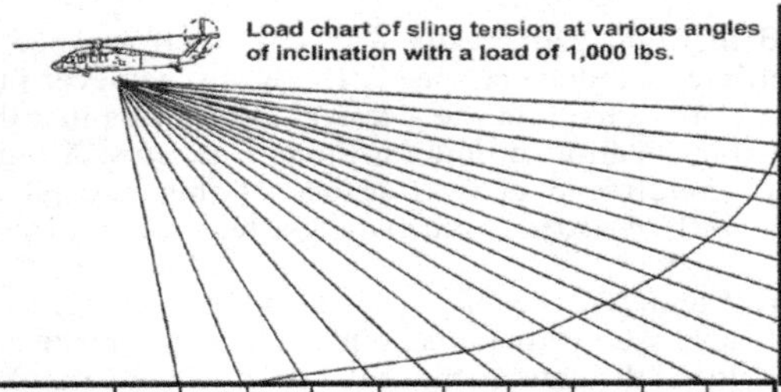

TOTAL VERTICAL LOAD (LB)	TOTAL SLING TENSION (LB)	ANGLE DEGREES
		0
1000	11473	5
1000	5759	10
1000	3863	15
1000	2924	20
1000	2366	25
1000	2000	30
1000	1743	35
1000	1555	40
1000	1414	45

TOTAL VERTICAL LOAD (LB)	1000	1000	1000	1000	1000	1000	1000	1000	1000	1000
TOTAL SLING LOAD (LB)	1000	1003	1015	1035	1064	1103	1154	1220	1305	1414
ANGLE DEGREES	90	85	80	75	70	65	60	55	50	45

To compute the tension in each leg of a sling assembly, divide the total sling tension by the number of sling legs. Divide the load by 1,000 and multiply by the total sling tension found in the table for the corresponding angle.

Figure 5-16. Load chart of sling tension at various angles of inclination with a load of 1,000 pounds.

5-10. HOOKUP AND RELEASE PROCEDURES

Hooking up a load requires a team effort. The signalman must position the aircraft over the load. He does this so the slingload team can discharge the static electricity and attach the load to the aircraft as quickly and safely as possible. Most of the time, the air crew releases the load. This seldom requires any ground crew except the signalman.

a. **Ground Crew Protective Measures and Equipment**. Working around hovering helicopters exposes ground crews to a variety of dangers. Leaders must do everything they can to ensure the safety of the ground crews. The crews themselves should use the following safety equipment (Figure 5-17):

(1) ***Helmet***. This protects the wearer from head injuries caused by flying debris. It also protects him if his head were to get caught between the aircraft and the load, for example. Wearers must keep helmets securely fastened.

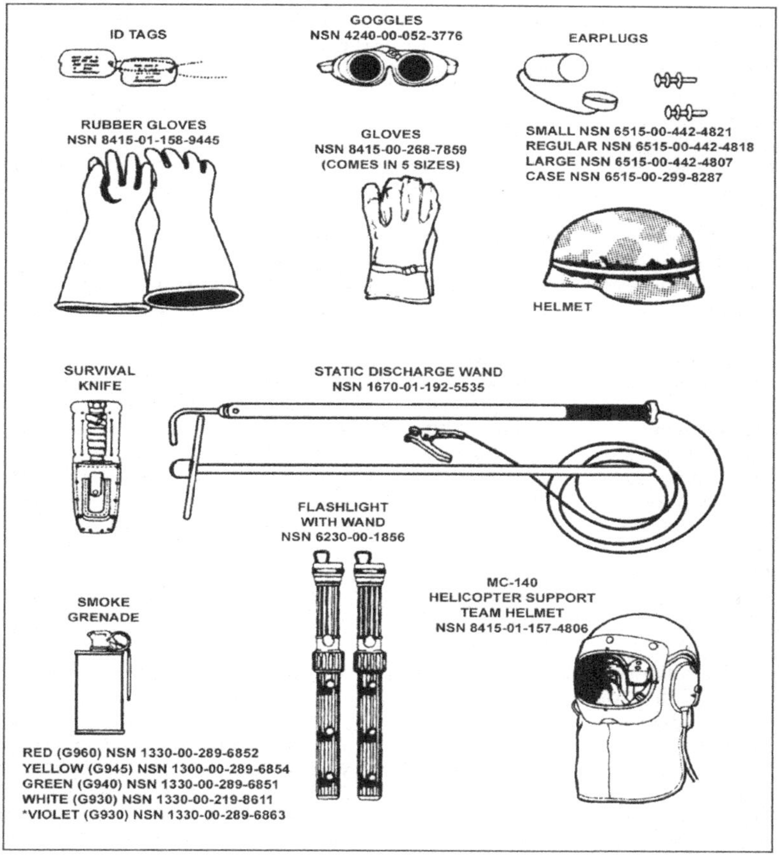

Figure 5-17. Protective equipment.

(2) ***Protective Mask or Dust Goggles with Respirator***. In high dust or debris environments, each crewmember wears a mask or goggles. This protects the crewmembers' faces, eyes, and respiratory systems from the airborne particles stirred up by the rotor wash. The mask protects better but can cause problems with depth perception (important for signalmen).

(3) ***Earplugs or Suitable Substitute***. These protect against the excessive noise associated with hovering aircraft. They also prevent debris from entering the ear canal.

(4) ***Hand Protection***. Marine Corps and Navy personnel must wear gloves designed to protect electrical workers against burns due to static discharges. Everyone else (USA,

USAF, and USCG) should wear leather gloves to help protect the hands and fingers. Whoever must use the static wand to discharge static electricity should wear gloves designed for adequate protection from static discharge burns.

(a) Inspect shockproof gloves before and after each operation. Check for excessive wear, fraying, holes, and tears. Do not use a torn glove. Even a small hole leaves a person unprotected from static electric shock.

(b) Check each glove for holes by filling it with water and squeezing it while holding the open end closed. Or, blow air into it like a balloon and submerge it in water. Any holes will cause air bubbles.

(5) ***Static Discharge Wand***. The static wand protects the hookup man from static electric shock by grounding the cargo hook. In flight, the stored static electric energy of the helicopter increases with its weight, with low humidity, and with the amount of debris blown around by the rotor system (dust, sand, or snow). Thunderstorms can cause huge discharges of static electricity. When the helicopter lands and touches the ground, this charge grounds out. While the helicopter flies or hovers to make a slingload drop, the charge remains stored in the aircraft. As soon as the ground crew member connects the apex fitting to the cargo hook, he provides a path for a charge to follow into the ground. This charge can cause severe electrical burn or injury.

(a) To avoid the possibility of a static electric shock, ground the cargo hook using grounding stakes and static discharge wands (field-expedient and manufactured). Since these wands connect the helicopter to the ground, the static electric charge dissipates. This protects the hookup man from receiving a shock when he connects the apex fitting to the cargo hook.

(b) Inspect the static discharge wand to make sure it is in serviceable condition. Drive the grounding stake opposite the ground crew's exit direction. This keeps them from tripping on the cable as they leave.

(c) Drive the stake into the ground until it seats firmly--at least 6 to 8 inches (firm ground) and 24 inches (sandy or loose soil). Drive the stake in at a 45-degree angle away from the side of the load, in case someone falls on it. Connect the cable clamp to the vertical shaft of the stake.

(d) When operating on concrete or asphalt surfaces, position the loads as close to the edge of the surface as you can. This allows you to drive the grounding stake into the ground.

(e) Do not hold the static discharge wand within 14 to 16 inches of the metal hook end--a strong static charge can jump as far as 12 inches. During the hookup operation, the static discharge wand must stay in contact with the cargo hook. If contact fails, all ground crew members must pull back from the hook until someone can reestablish contact between the wand and the aircraft's cargo hook.

(6) ***Other Equipment***. Use smoke grenades to mark the location of the landing site or to indicate wind direction. Use flashlights with wands to give arm-and-hand signals at night.

b. **Safety**. In addition to using the proper equipment, soldiers must also follow these other safety measures:

(1) Wear long-sleeved shirts with the sleeves rolled down and fastened. Button your shirt collar. Tuck shirttails or jacket bottoms into your trousers.

(2) Police the operational area thoroughly before conducting sling-load operations. This cuts down on the amount of debris thrown about by rotor wash.

(3) Stay alert during hookup and release operations; sound judgment and common sense hold the keys to success. Stay ready to get clear of the load. Soldiers have been crushed between the aircraft and loads. Some have had loads dragged over them; others took an unwanted ride when they somehow entangled themselves with the load. Whenever you have to make the hookup, take special care. Slings under tension can easily crush an arm or leg against the load. Some of the particular hazards associated with loads include--

(a) *Cargo Extensions or Projections*. Gun tubes, landing gear, missile launchers, bridge planks, and so forth can interfere with or injure you by striking or tripping you. Stand clear of such projections or position yourself so you can clear the load at once.

(b) *Sharp Projections, Hooks, Handles, Racks*. If possible, avoid these. Examples include protruding handles or levers such as tarpaulin tie-down hooks, door handles, spare-tire racks, and similar projections. Sharp edges can cause serious injury. You can quickly get hooked to the load if your clothing or equipment catches on something. Keep alert and ready to move out of danger immediately.

(c) *Top-Heavy or Narrow-Based Loads*. Treat with caution any top-heavy or narrow-based loads that the rotor wash could blow over. If possible, lay this kind of load on its side before starting the hookup. If you cannot do this, position the crew on the side or end of the load that is least likely to tip. Again, stay ready to move away from the danger quickly.

(d) *High Loads*. High loads can seriously injure you if you climb up on them to hook them up. Rotor wash can sweep you off, or you might have to jump to avoid a dangerous situation. Pay attention to where you stand. Try to stand on a lower projection or step rather than on top of the load. This way, if the aircraft makes contact with the load, it does not catch you in between. Also, try to work from a crouched position or from your hands and knees. Keep solid footholds and handholds, and stay ready to move quickly out of the way, if you need to. If possible, back a vehicle up to the load and use it as a working platform. (Move it out of the way before the aircraft starts to lift the load.)

c. **Ground Crew Emergency Conduct**. When an aircraft hovering over a slingload suffers an emergency severe enough for the pilot to have to set the aircraft down, he will do so. This can happen all at once such as in a controlled crash. For this reason, whenever an aircraft seems to be having trouble, all members of the ground crew should clear the slingload point by moving to a location coordinated with the aviation unit. Once they have moved far enough away, they should each take a prone position or seek cover until the aircraft lands. Two situations require special note:

(1) ***Signalman***. Face the aircraft; move to a safe spot.

(2) ***Hookup Men***. Try to work along the right side of the load. This way, you do not have to climb over or go around the load to seek safety. You can instead move directly off and away from the load. If the load is a heavy piece of equipment, you might want to keep the load between you and the aircraft while you are moving. This offers you some protection if the aircraft were to crash.

d. **Ground Crew Duties**. Normally, the ground crew consists of one signalman and two hookup men, with one hookup man acting as static wand man.

(1) ***Duties of the Signalman***.

(a) *Before the Aircraft Arrive(s)*. Direct the positioning of the load. Supervise the inspection of the load for proper routing of the slings and proper preparation. Ensure that the load is ready to fly.

(b) *As the Helicopter(s) Approaches*. Station yourself 20 meters in front of the load where you can best maintain eye contact with the crew. Give the arm-and-hand signal of "assume guidance." As the helicopter nears the load, use arm-and-hand signals to position the cargo hook directly over the load, close enough that the hookup men can place the apex fitting onto the cargo hook. During this time, position yourself so the pilot can see your signals easily. Since the pilot of an Army aircraft sits on the right side of the aircraft, you will usually stand just to the right of the aircraft. If the terrain forces you to stand somewhere else, make sure the pilot can see you at all times.

(c) *During the Hookup Process*. Watch the cargo hook and apex fitting. After hookup, the pilot hovers the aircraft until the hookup men clear away from the load. When they have moved clear, you will signal the aircraft upward slowly, so the sling legs gradually take up the load. You must do this to make sure the sling legs *clear* the load. If the sling legs foul, motion the pilot downward, and then instruct him to cut away the load. If you did a good job of hooking up the load, and if the load suspends properly below the aircraft, then give the aircraft the signal to depart. Then, move quickly aside to clear the helicopter's path.

DANGER

AT NO TIME WILL THE SIGNALMAN OR ANY OTHER MEMBER OF THE SLINGLOAD TEAM ALLOW A SUSPENDED LOAD TO PASS OVER HIS HEAD.

(2) ***Duties of the Hookup Men***.

(a) One of you handles the static discharge wand and the cargo hook. The other controls the apex fitting of the slingload. Together, you must complete the hookup fast to reduce helicopter hover time and to reduce your exposure time under the helicopter.

(b) Position yourselves by the load so that, while the helicopter hovers over the load, you can quickly complete the hookup (Figure 5-18). You must also make sure the signalman can continually observe the operation.

(c) When the helicopter moves into the correct position for hookup, whichever one of you is the static wand man must ground the aircraft. Touch the static wand to the cargo hook (Figure 5-19) and keep it there to maintain a continuous ground.

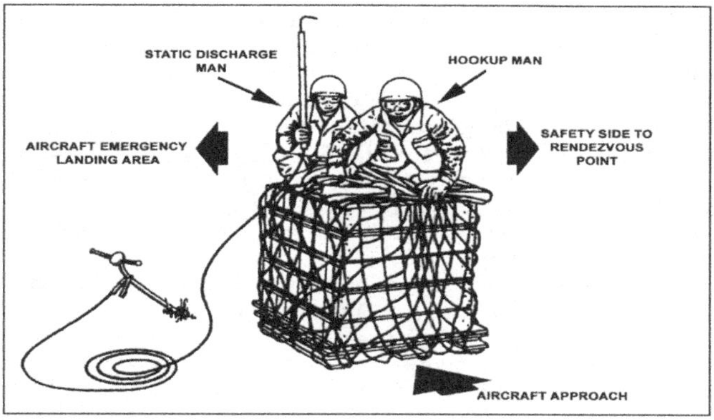

Figure 5-18. Position of hookup team.

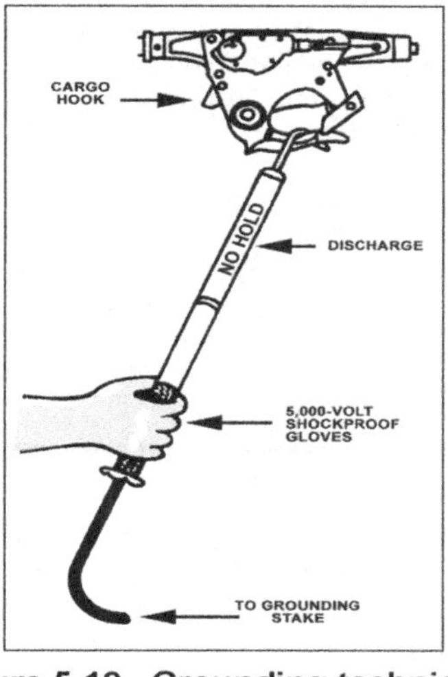

Figure 5-19. Grounding technique.

(d) Once the static wand man grounds the aircraft, the other hookup man places the apex fitting onto the cargo hook, then checks to make sure that the hook is properly closed (and locked, if required).

(e) After you properly hook up the load to the aircraft, both of you must move quickly aside to the location coordinated with the aviation unit. If the signalman learns that any of the legs have fouled, he notifies the pilot at once. Also, you will have to rehook the load.

e. **Release Procedures**. For this mission, leaders refer to the hookup men as the "cargo release team." As the helicopter approaches the site, the pilot takes instructions from the signalman, who guides the aircraft into position for cargo release. The cargo release team stands by, unless it must release the load manually. The signalman directs the aircraft to set the load on the ground. He gives the release signal. At this time, the apex fitting should fall free of the cargo hook. If it does not, the signalman has the aircraft hover, then he directs the cargo-release team to move under the helicopter and manually release the load from the hook. The load clears the hook. After the release, everyone moves out from under the aircraft. The signalman directs the aircraft to depart and quickly moves out of its path.

(1) If the pilot cannot activate the cargo hook from within the helicopter, and if cargo release personnel open it, then ground crew members must use emergency cargo release procedures:

(a) Try to disassemble the doughnut.

(b) Pass the ADS through the hook.

(2) If the cargo hook is attached to a clevis or apex fitting, unscrew the nut on the clevis or fitting and remove the pin.

(3) If necessary, derig the load so the aircraft can set down.

f. **Hookup Procedures During Whiteouts or Brownouts**. The hazards of these conditions (snow or dust) prevent the use of a signalman or a hovering hookup.

(1) Rig the load with a 20-foot or a 40-foot extension (as required) using 20-foot ADS with two or three loops and the appropriate number of Type IV link assemblies. Place an apex fitting at the end of the extension.

(2) Lay the extension to the left of the load. The aircraft approaches normally, then taxis to the location of the apex fitting and stops. Once the aircraft lands, the hookup person moves to the aircraft and attaches the apex fitting to the cargo hook. The aircraft suspends the load and departs as directed by the GTA.

(3) When attaching the extension to skid-equipped helicopters, such as the UH-1H, take care that the sling goes forward of *but not through* the skid. Then attach it to the cargo hook.

WARNING

Before the operation begins, coordinate the ground crew's evacuation route to a rendezvous point. Proper coordination with the liaison officer or helicopter crew prevents confusion.

Helicopter emergency procedures depend on terrain, wind direction, and pilot choice.

Good coordination prevents the helicopter and ground crews from moving in the same direction.

5-11. SLINGLOAD INSPECTION RECORD

To improve slingload safety, the Department of the Army implemented inspection procedures for all ARMY equipment moved by the slingload method of air delivery. These procedures went into effect 1 October 1997. All Army loads require inspection by a qualified inspector before the arrival of the supporting aircraft. The inspector completes the *Sling Load Inspection Record.* (Figure 5-20, page 5-26, shows an example completed DA Form 7382-R. FM 10-450-3.) (Appendix I discusses slingload inspection in greater detail.)

a. **Inspector Qualifications**. Inspectors must hold the grade of E-4 or more. They must also either be a Pathfinder, a certified sling load inspector, or qualified in air assault.

b. **Distribution of the Sling Load Inspection Record**. Reproduce the slingload inspection record onto 8 1/2 by 11-inch paper, get it through official distribution channels, or downloaded from the AEL. Complete the inspection record in triplicate. Copies of the completed form are distributed as follows:

(1) Give a copy to the supporting aviation unit.

(2) Securely tape or tie a copy to the load.

(3) Give a copy to the supported unit.

SLING LOAD INSPECTION RECORD

For use of this form, see FM 10-450-3; the proponent agency is TRADOC

1. SUPPORTED UNIT	2. ITEM DESCRIPTION AND SERIAL/BUMPER NO.	3. WEIGHT
HHC 1-507 PFDR	CARGO NET	2,000 LBS

4. SUPPORTING AVIATION UNIT	5. TYPE AIRCRAFT	6. RIGGED IAW FM NO
DOT	CH-60A	FM 10-450-3

SAMPLE

INITIAL ONLY ITEMS APPLICABLE TO YOUR SPECIFIC LOAD	LOAD RIGGED BY	LOAD INSPECTED BY
7. VEHICLE OR LOAD		
A. CORRECTLY POSITIONED	X	X
B. EMERGENCY BRAKE SERVICEABLE AND SET	X	X
C. FUEL LEVEL NOT TO EXCEED 3/4 TANK	X	X
D. PREPARED AND PADDED IAW THE APPROPRIATE FM	X	X
8. SLING SET		
A. CORRECT NUMBER AND SIZE (10K OR 25K)	X	X
B. INSPECTED FOR SERVICEABILITY IAW FM 10-450-3	X	X
C. SLING LEGS PROPERLY ROUTED AND ATTACHED TO LIFT POINTS	X	X
D. CORRECT LINK COUNT FRONT AND REAR	X	X
E. CHAIN SECURED IN GRAB LINK	X	X
F. EXCESS CHAIN TIED OR TAPED (10 LINKS OR MORE)	X	X
G. BREAKAWAY TIES INSTALLED	X	X
H. APEX ATTACHED	X	X
I. APEX SPACER INSTALLED IF REQUIRED	X	X
J. REACH PENDANT INSTALLED IF REQUIRED	X	X
9. A-22 CARGO BAG		
A. INSPECTED FOR SERVICEABILITY IAW FM 10-450-3	X	X
B. RIGGED IAW FM 10-450-3	X	X
C. SUSPENSION WEBS ATTACHED TO CONTAINER AND TAPED	X	X
D. CLEVIS BOLT THROUGH SUSPENSION WEB D-RINGS (4 EACH)	X	X
E. CORRECT SLING LEG ATTACHED	X	X
F. APEX ATTACHED (CORRECT TYPE)	X	X
G. APEX SPACER INSTALLED IF REQUIRED	X	X
10. CARGO NETS	5K	
A. CORRECT SIZE (5K OR 10K)	JD	MC
B. INSPECTED FOR SERVICEABILITY IAW FM 10-450-3	JD	MC
C. LOAD CORRECTLY POSITIONED	JD	MC
D. LIFTING LEGS PROPERLY CONNECTED TO APEX FITTING	JD	MC
E. HOOKS TAPED	JD	MC
F. LIFTING LEGS TAPED OR TIED (BREAKAWAY)	JD	MC
G. CORRECT NUMBER AND SIZE SLING LEGS	JD	MC

REMARKS

11. LOAD RIGGED BY:

UNIT (PRINT)	NAME (PRINT)	INITIALS	RANK
HHC 1-507 PFDR	JOHN DOE	JD	SSG
SIGNATURE: John Doe		DATE:	

12. LOAD INSPECTED BY:

UNIT (PRINT)	NAME (PRINT)	INITIALS	RANK
HHH 1-507	MICHAEL CLARK	MC	SFC
SIGNATURE: Michael Clark		DATE: DEC 4 '01	

DA FORM 7382-R, MAR 97 USAPPC V1.00

Figure 5-20. Example completed DA Form 7382-R.

THIS CHAPTER IMPLEMENTS STANAG 3570.

CHAPTER 6
DROP ZONES

The ground unit commander designates the drop zone, usually with the DZST leader's technical help. The drop zone is where drop aircraft deliver personnel and equipment by parachute or free drop. The commander selects a DZ location that best supports the tactical plan. In the case of tactical training, the commander checks the USAF assault zone availability report (AZAR) to see if an approved DZ already exists within the tactical area. If the AZAR does not include a DZ in that area, the commander must assess the tactical situation before choosing a DZ location.

Section I. SELECTION FACTORS

The commander uses the drop zone selection factors discussed in this section to analyze the suitability of a drop zone.

6-1. AIRDROP AIRSPEEDS

The speed of the aircraft determines how long the aircraft will remain over the drop zone. Table 6-1 provides the fixed-wing and rotary-wing airdrop speeds in knots indicated airspeed (KIAS).

TYPE OF AIRCRAFT	DROP SPEED
UH-1	50 to 70 knots (Best speed is 70 knots)
UH-60	65 to 75 knots (Best speed is 70 knots)
CH-46 (USMC)	80 knots
CH-47	80 to 110 knots (Best speed is 90 knots)
CH-53	90 to 110 knots
CH/HH3 (USAF)	70 to 90 knots
C-5/130/141/KC-130	130 to 135 knots (personnel)
C-5/130/141/KC-130	130 to 150 knots (Door bundles, CDS, and heavy equipment--best speed for all loads is 130 knots)
C130	140 knots (CDS)
C5/17/141/KG130	150 knots (CDS)

Table 6-1. Airdrop speeds.

6-2. DROP ALTITUDE

The DZ leader measures drop altitude (Table 6-2, page 6-2) from the highest point on the DZ (the highest field elevation) to the aircraft. In combat (wartime) operations, airborne and airlift commanders jointly determine drop altitudes.

a. Table 6-3, page 6-3, shows drop altitudes for different types of training missions. (For more information on drop altitudes, see AFI 11-231 and AFI 11-410.)

b. The aircraft altimeter displays altitude in feet above sea level, not in feet above the highest point on the ground. Thus, the pilot might request the drop altitude in "feet indicated," that is, the number that should appear on the altimeter at the time of the drop. You can calculate this simply by following this example:

(1) Obtain the drop altitude, that is, the distance in feet from the highest point on the drop zone (field elevation) to the desired altitude of the aircraft. In this example, drop altitude equals 800 feet (A, Table 6-2).

(2) Obtain the highest field elevation in feet above sea level. Round this number up to the nearest multiple of 50 (round 537 up to 550, for example) (B, Table 6-2). For purposes of obtaining the drop altitude in feet indicated, use this number for field elevation.

(3) Sum the two numbers you obtained to yield drop altitude in feet indicated (C, Table 6-2).

A	Distance from highest field elevation in DZ to aircraft, in feet.	800 Feet AGL
B	Highest point on DZ--highest field elevation in feet above sea level, round up to next 50, for example, round 505 up to 550.	+ 550 Feet Field Elevation
C	Drop altitude in feet indicated	1,350 Feet Indicated

Table 6-2. Example calculation of drop altitude in feet indicated.

6-3. ESTIMATION OF DROP ZONE TIME REQUIREMENT

For personnel, allow one second for each jumper after the first. For example, ten jumpers minus the first jumper equals nine jumpers. Multiply nine times one second. Allow nine seconds for all ten jumpers to get out the door. For equipment, allow three seconds for each door bundle after the first. For example, five bundles minus the first bundle equals four bundles. Multiply four times three seconds each. Allow twelve seconds to get the equipment out the door.

6-4. METHODS OF DELIVERY

When considering the method of delivery, pathfinders must take extra care during high-velocity airdrops or free drops around built-up areas or airfields. Either could damage buildings or airstrips.

a. **Low-Velocity Airdrop**. Pilots use a low-velocity airdrop for sensitive equipment and for personnel. The parachute slows the rate of the descent, allowing a soft landing.

b. **High-Velocity Airdrop**. Pilots use a high-velocity airdrop to deliver certain supply items. The load must be rigged in an airdrop container with an energy dissipater attached to its underside and a ring-slot parachute attached to the top. The chute stabilizes the load and reduces the rate of fall, ensuring an acceptable landing shock.

c. **Free Drop**. Pilots use free drop for nonsensitive items only. This type of load has no parachute to stabilize or slow its rate of descent.

PERSONNEL		
Tactical Training		800 FT
Basic Airborne Trainees		1,250 FT
HALO (Minimum Opening)		2,500 FT
SATB/TTB (Use stand drop altitude of simulated load)		500 FT
Door bundles G13, G14 or T10		300 FT
CONTAINER DELIVERY SYSTEM		
G-12D:	1 to 6 bundles, single parachute	400 FT
	7 or more bundles, single parachute	600 FT
	Cluster of 2 parachutes	600 FT
G-12/	Single parachute	400 FT
13E:	Cluster of 2	550 FT
G-13:	Single parachute or cluster of 2	400 FT
	Cluster of 3	500 FT
G-14:	Single parachute or cluster of 2	300 FT
	Cluster of 3	400 FT
12-Foot High Velocity		400 FT
26-Foot High Velocity		500 FT
SATB-C/TTB		See parachute load to be simulated
HEAVY EQUIPMENT		
G-12D		650 FT
G-11A:	Single parachute	900 FT
	Cluster of 2 to 7	1,100 FT
	Cluster of 8	1,300 FT
G-11B:	Single parachute	700 FT
	Cluster of 2 to 4	750 FT
	Cluster of 5 to 7	1,100 FT
	Cluster of 8	1,300 FT
G-11C:	Cluster of 8	1,300 FT
SATB-H/TTB		See parachute load to be simulated.
NOTES:	1. Minimum drop altitude for heavy equipment using 5,000-pound parachute release system is 1,000 feet above ground level.	
	2. Combination drops will use the highest airdrop altitude.	
ROTARY WING AERIAL DELIVERY		
Personnel:	Day or night	1,500 FT
Bundles:	Day	300 FT
	Night	500 FT

Table 6-3. Aerial delivery altitudes.

d. **Added Risk**. When determining the suitability of the DZ and considering the method of delivery around built-up areas or airfields, the pilot also considers the added risk of damage to buildings when using high-velocity or free-drop methods.

6-5. OBSTACLES

To ensure a safe airdrop, and to make sure soldiers can recover and employ airdropped personnel and equipment, the pathfinders should clear the DZ and adjacent areas of all obstacles.

a. **Obstacles to Personnel**. This includes anything, natural or manmade, that could harm a parachutist.

b. **Obstacles to Equipment**. This includes anything, natural or manmade, that could hinder the recovery of equipment:

(1) ***Trees***. Trees 35 feet or higher, which would impede recovery of personnel or equipment.

(2) ***Water***. Water more than 4 feet deep within 1,000 meters of any edge of the drop zone presents an obstacle.

(3) ***Power Lines***. Any power lines that carry an active current of 50 volts or greater present a real and dangerous obstacle. Based on the reported altitude wind, pathfinders create a safety zone for parachutists by turning off all power within that zone. This helps prevent injuries. Figure 6-1 shows the recommended sizes of the safety zone for various altitude wind velocities during static line operations.

(4) ***Other***. Any other conditions, such as barbed wire fences, swamps, rocks, ditches, and gullies, that could injure parachutists or damage equipment.

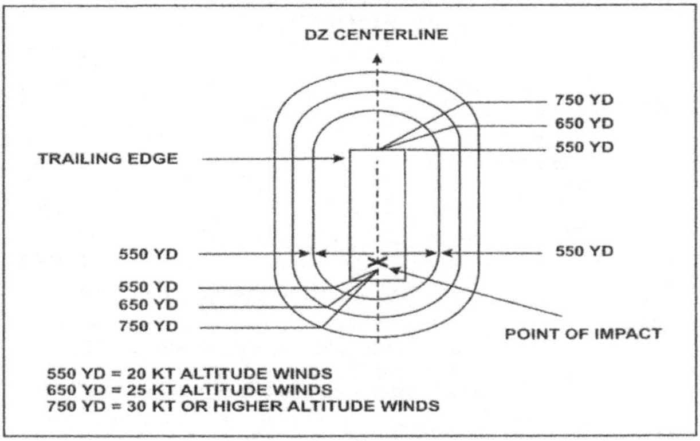

Figure 6-1. Recommended safety zones for high-tension lines.

6-6. ACCESS

Pathfinders should avoid any DZ that has a major obstacle between it and the objective area. Pathfinders should also make sure the area has adequate routes to conduct troop movement and to recover equipment.

6-7. SIZE

USAF doctrine determines the size of computed air release point DZs during peacetime drop operations.

a. **One Jumper**. The CARP DZ size for one jumper is at least 549 meters (600 yards) by 549 meters (600 yards). For each additional jumper, add 64 meters (75 yards) to the length of the DZ.

b. **Container Delivery System**. Table 6-4 shows the CARP DZ sizes for the container delivery system.

CDS (C-130)				
ALTITUDE (AGL) IN FEET	WIDTH	NUMBER OF CONTAINERS: SINGLE	NUMBER OF CONTAINERS: DOUBLE	LENGTH
To 600	400 Yards	1	1 to 2	400 Yards
		2	3 to 4	450 Yards
		3	5 to 6	500 Yards
		4	7 to 8	550 Yards
		5 to 8	9 or more	700 Yards
Above 600	Add 40 yards to DZ width and length for each 100 feet above 600 feet (Add 20 yards to each side of the DZ).			

CDS (C-141)				
ALTITUDE (AGL) IN FEET	WIDTH	NUMBER OF CONTAINERS: SINGLE	NUMBER OF CONTAINERS: DOUBLE	LENGTH
To 600	450 Yards	1	1 to 2	590 Yards
		2	3 to 4	615 Yards
		3	5 to 6	665 Yards
		4 to 8	7 to 16	765 Yards
		9 to 14	17 to 28	915 Yards
		15 to 20	29 to 40	1,065 Yards
Above 600	Add 40 yards to DZ width and length for each 100 feet above 600 feet. Add 20 yards to each side of the DZ.			

Table 6-4. Size criteria for tactical airlift drop zones, CDS.

c. **Personnel**. Table 6-5, page 6-7 shows the CARP DZ size for personnel. AFI 13-217, 1 Jun 99, provides additional size requirements.

d. **Heavy Equipment**. For this, make sure the CARP DZ size (Table 6-5, page 6-7) measures at least 600 yards (549 meters) wide by 1,000 yards (915 meters) long for one platform.

- C-130--add 366 meters (400 yards) to the length for each additional platform.
- C-141--add 458 meters (500 yards) to the length for each additional platform.

e. **More Than One Aircraft Outside Trail Formation**. In this situation, add 92 meters (100 yards) to the width of all CARP DZs (46 meters--50 yards--each side).

f. **Official Sunset to Sunrise**. During these times, add 100 yards (92 meters) to the length and width of all CARP DZs--50 yards (46 meters) each side and to each end.

NOTE:
- To convert yards to meters, multiply yards by 0.9144.
- To convert meters to yards, divide meters by 0.9144.

g. **Army VIRS Drop Zone**. For this, allow a minimum size of 300 yards by 300 yards (275 meters by 275 meters). To determine the required size of Army VIRS DZs, use the D=RT formula. For personnel jumps, allow a 100-meter buffer zone at the leading and trail edges of the DZ. If local regulations permit, the local commander can waive these buffer zones.

h. **Ground-Marking Release System Drop Zone.** Again, allow a minimum size of 300 yards by 300 yards (275 meters by 275 meters). Use the D=RT formula (Figure 6-2, page 6-8) to determine other size requirements. In the minimum size, include the 100-meter buffer (personnel drops only) at both the leading and trail edges of the DZ. (See Table 6-1, page 6-1, for airdrop speeds.)

i. **Parachutists or Bundles**. To calculate the number of either parachutists or bundles that a GMRS DZ of given length can accept in one pass, use the T = D/R formula (Figure 6-3, page 6-9). You must know the type of aircraft.

j. **Load Drift Under Parachute**. To calculate the amount of drift experienced by a load under a parachute, use the D = KAV formula (Figure 6-4, page 6-10). Always round up to the next whole number.

k. **Wind**. Measuring wind on the drop zone entails measuring both surface wind and mean effective wind.

(1) Use the AN/PMQ-3A, or any other acceptable wind measuring device, to measure surface (ground) wind speed, especially for personnel and heavy equipment operations.

(2) Mean effective wind (MEW) refers to the average wind from ground level to drop altitude. Measure the magnetic azimuth to the balloon and note the reciprocal heading. This gives you the MEW direction to report. Use the PILOT balloon (PIBAL) to measure MEW. PIBAL circumferences include--

- 10 grams for day--57 inches.
- 30 grams for day--78 inches.
- 10 grams for night--74 inches.
- 30 grams for night--94 inches.

(3) At night, attach a small, liquid-activated light to the balloon to aid in observation. For PIBAL charts, see Table 6-6A, page 6-11 (10-gram helium balloon) and Table 6-6B, page 6-12 (30-gram helium balloon).

l. **Forward Throw**. This refers to the effect of inertia on a falling object. An object that leaves an aircraft moves at the same speed as the aircraft. The parachutist (or bundle) continues to move in the direction of flight until the dynamics of gravity and the parachute take effect. Forward throw for rotary-wing aircraft equals half the aircraft speed, expressed in meters. Table 6-7, page 6-12 shows the amount of forward throw from a fixed-wing aircraft, by distance.

6-8. APPROACH AND DEPARTURE ROUTES

Pathfinders must choose adequate routes for the aircraft to and from the DZs. To ensure this, they consider--

- Enemy situation and location.
- Obstacles to the aircraft (TV towers, high-tension lines).
- Terrain higher than the DZ.
- No-fly areas.

HEAVY EQUIPMENT			
ALTITUDE (AGL IN FEET)	WIDTH	DROP ZONE LENGTH	
		ONE PLATFORM	ADDITIONAL PLATFORMS
To 1,100	600 Yards	1,000 Yards	Add 400 yards (C-130) or 500 yards (C-141 or C-5) to trailing edge for each additional platform.
Above 1,100	Add 30 yards to DZ width and length for each 100 feet above 1,000 feet (Add 15 yards to each side of the DZ).		
PERSONNEL			
ALTITUDE (AGL IN FEET)	WIDTH	DROP ZONE LENGTH	
		ONE PLATFORM	ADDITIONAL PLATFORMS
To 1,000	600 Yards	600 Yards	Add 75 yards to trailing edge for each additional parachutist. When using CAPES, add 100 yards each, instead.
Above 1,000	Add 30 yards to DZ width and length for each 100 feet above 1,000 feet (Add 15 yards to each side of the DZ).		

NOTES:
1. For day visual formations, increase width by 100 yards (50 yards each side).
   For SKE formation, increase width by 400 yards (200 yards each side).
   Official sunset to sunrise, increase width by 100 yards for single ship visual drops (50 yards each side) or 200 yards for visual formations (100 yards each side).
2. Official sunset to sunrise, increase length by 100 yards for visual drops (50 yards each end).

Table 6-5. Size criteria for tactical airlift drop zones, personnel, and heavy equipment.

D = RT

Determine additional size requirements.

D = Length of DZ in meters.

R = Aircraft's rate of speed in meters per second (MPS). To convert knots to meters per second--

- Multiply the number of knots by 0.51.
- Do not round this number up or down.

T = Time required to exit each load:

- Parachutists (personnel) require 1 second each, after the first, which is free. The formula for computing the total seconds required to drop personnel is (N - 1) x 1, with N equal to the total number of personnel. Thus, ten personnel require--

  (10 – 1) X 1 = 9 X 1 = 9 seconds

- Door bundles require 3 seconds each, after the first bundle. The formula for computing the total seconds required to drop bundles is (N - 1) x 3, with N equal to the total number of bundles. Thus, three bundles require--

  (3 – 1) X 3 = 2 X 3 = 6 seconds

EXAMPLE PROBLEM

What is the minimum GMRS DZ length needed for a C-130 to drop ten parachutists?

D = Length of DZ in meters (unknown).

R = 66.30 meters per second (130 x 0.51).

T = 9 seconds (1 second per parachutist, not counting the first).

SOLUTION

D = R x T

D = 66.3 MPS X 9 seconds = 596.7 meters

(Round up to nearest whole number, which is 597.)

D = 597 meters of usable DZ required.

For personnel drop zones, add a 200-meter buffer--100 meters on the leading edge and 100 meters on the trailing edge--to total 797 meters for the drop zone. (The commander can waive this requirement.)

Figure 6-2. Example usage of D = RT formula.

T = D / R

Calculate number of parachutists or bundles that a GMRS DZ of given length can accept in one pass

T = Amount of time in seconds that the aircraft will be over the DZ.

D = Distance of DZ in meters (length).

R = Rate of aircraft's speed expressed in meters per second. To convert knots to meters per second, multiply the knots by .51. Round this number up to the nearest whole number.

EXAMPLE PROBLEM

How many parachutists from a C-130 can a 750-meter-long GMRS drop zone accept on each pass?

T = The number of parachutists that can drop on a 750-meter long GMRS DZ.

D = 550 meters (750-meter drop zone less 100-meter buffer at each end).

R = 130 x 0.51 = 66.3 rounded up to 67.

SOLUTION

T = **D/R**

D/R = (550/67) = 8.2

T = = 8 seconds (always round down).

DZ Can accept nine parachutists per pass: eight parachutists, at one each second, plus one free.

Figure 6-3. Example usage of T = D/R formula.

Section II. DROP ZONE SUPPORT TEAM

The drop zone support team plans, establishes, and operates day and night drop zones for personnel and resupply missions flown by fixed-wing and rotary-wing aircraft. The DZST is responsible for accomplishing the mission on the DZ. In operations without the USAF special tactics team, the DZST will shoulder the overall responsibility for the conduct of operations on the DZ. The DZST represents both the airborne and airlift commanders. The DZST leader assumes all the responsibilities normally associated with the USAF STT and the DZSO.

6-9. ORGANIZATION

The DZST must have at least two members. It might need more, depending on the complexity of the mission. However, additional team members do not need DZST qualification. The senior member of the DZST functions as the team leader. He must hold the rank of NCO (sergeant or above in the US Army, E-4 or above in the USAF or USMC), an officer, or the civilian equivalent. He must have completed the appropriate initial training as a DZST member and must satisfy current parent service requirements. To conduct personnel and heavy equipment drops, he must also hold current jumpmaster qualification.

D = KAV

Calculate drift

D = The amount of drift in meters.
K = Load drift constants:

- Personnel .... 3.0 meters
- Bundle .... 1.5 meters
- Equipment .... 1.5 meters
- Containerized delivery system (CDS) .... 1.5 meters
- Tactical training bundle .... 2.4 meters

A = Express drop altitude in hundreds of feet: For 800 feet, say "8"; for 850, say "8.5."
V = Velocity of the wind (Use either surface wind speed or MEW).

EXAMPLE PROBLEM

For a drop altitude of 800 feet and a wind speed of 11 knots, calculate a jumper's drift:

K = Load drift constant for jumper (personnel) = 3 meters.
A = Drop altitude = 800 feet, so in this example, A = 800/100 = 8.
V = Wind speed = 11 knots.

SOLUTION

D = **K x A x V**
D = 3 x 8 x 11
D = 264, in the example conditions, a jumper drifts 264 meters.

Figure 6-4. Example usage of D = KAV formula.

6-10. MISSIONS

Primary missions of the DZST include wartime CDS drops to battalion or smaller units. They also make peacetime, visual, meteorological-condition drops, with one to three aircraft, for personnel, CDS, and heavy equipment. Secondary missions include wartime drops of brigade size or larger units, peacetime drops of C-130 all-weather aerial delivery system (AWADS) involving one to three aircraft, or VMC drops of four or more aircraft.

6-11. EQUIPMENT FAMILIARIZATION

The DZST leader must know how to use equipment to set up, mark, and operate the drop zone. This includes the following items, more or less, depending on the mission:

a. **Anemometer**. The handheld, calibrated AN/PMQ 3A (NSN: 6660-00-515-4339) measures surface wind. With the trigger pressed down, the correctly oriented anemometer gives wind direction in degrees. It can read the wind from 0 to 15 knots on the low scale, and from 0 to 60 knots on the high scale. The anemometer requires recalibration every six months. Regardless of the method or device used to measure the wind on the DZ, the airborne commander must prevent static line personnel airdrops when the surface wind on the DZ exceeds 13 knots. The following USAIS messages authorize several commercially available anemometers for use in drop zone operations:

(1) DTG 101000Z Mar 94, Subject: Use of Anemometers During Airdrop Operations.

(2) DTG 212000Z Oct 94, Subject: Use of Turbo Meters During Static Line Airdrop Operations.

10-GRAM HELIUM BALLOON

INFLATE BALLOON TO 57" CIRCUMFERENCE FOR DAY AND 74" CIRCUMFERENCE FOR NIGHT.

DROP ALTITUDE IN FEET

ELEVATION ANGLE	500	750	1000	1250	1500	1750	2000	2500	3000	3500	4000	4500	ASCENSION TABLE	
70	02	02	01	01	01	01	01	01	01	01	01	01		
60	03	02	02	02	02	02	02	02	02	02	02	02		
55	03	03	03	03	03	03	03	03	03	03	03	03	TIME	ALT (FT)
50	04	04	03	03	03	03	03	03	03	03	03	03	0:10	80
45	05	04	04	04	04	04	04	04	04	04	04	04	0:20	170
40	06	05	05	05	05	05	05	04	04	04	04	04	0:30	250
35	07	06	06	06	06	06	06	05	05	05	05	05	0:40	330
30	08	07	07	07	07	07	07	07	06	06	06	06	0:50	400
25	10	09	09	09	08	08	08	08	08	08	08	08	1:02	500
24	11	10	09	09	09	09	08	08	08	08	08	08	1:10	540
23	11	10	10	09	09	09	09	08	08	08	08	08	1:20	610
22	12	11	10	10	10	10	09	09	09	09	09	09	1:30	670
21	12	11	11	10	10	10	10	10	10	10	10	10	1:43	750
20	13	12	11	11	11	11	11	10	10	10	10	10	1:50	790
19	14	13	12	12	11	11	11	11	11	11	11	11	2:25	1000
18	15	13	13	12	12	12	12	12	11	11	11	11	2:44	1100
17	16	14	13	13	13	13	12	12	12	12	12	12	3:05	1250
16	17	15	14	14	14	13	13	13	13	13	13	13	3:49	1500
15	18	16	15	15	14	14	14	14	14	14	14	14	4:30	1750
14	19	17	16	16	16	15	15	15	15	15	15	15	5:11	2000
13	21	19	18	17	17	17	17	16	16	16	16	16	6:34	2500
12	22	20	19	19	18	18	18	18	17	17	17	17	7:58	3000
11	24	22	21	21	20	20	20	19	19	19	19	19	9:22	3500
10	27	25	23	23	22	22	22	21	21	21	21	21	10:44	4000
09	30	27	26	26	25	24	24	24	23	23	23	23	12:08	4500

Table 6-6A. Conversion chart for 10-gram helium (pilot) balloon.

30-GRAM HELIUM BALLOON

INFLATE BALLOON TO 78" CIRCUMFERENCE FOR DAY AND 94" CIRCUMFERENCE FOR NIGHT.

DROP ALTITUDE IN FEET

ELEVATION ANGLE	500	750	1000	1250	1500	1750	2000	2500	3000	3500	4000	4500
80	01	01	01	01	01	01	01	01	01	01	01	01
70	03	03	03	02	02	02	02	02	02	02	02	02
60	04	04	04	04	04	04	04	04	04	04	04	04
55	05	05	05	05	05	05	05	05	05	05	04	04
50	06	06	06	06	06	06	06	06	05	05	05	05
45	07	07	07	07	07	07	07	07	07	06	06	06
40	09	08	08	08	08	08	08	08	08	08	08	08
35	10	10	10	10	10	10	10	09	09	09	09	09
30	12	12	12	12	12	12	12	11	11	11	11	11
25	15	15	15	15	15	15	14	14	14	14	14	14
24	16	16	15	15	15	15	15	15	15	15	15	15
23	17	17	16	16	15	15	15	15	15	15	15	15
22	18	18	17	17	17	17	17	16	16	16	16	16
21	19	19	18	18	18	17	17	17	17	17	17	17
20	20	20	19	19	19	19	18	18	18	18	18	17
19	21	20	20	20	20	20	19	19	19	19	19	18
18	22	22	21	21	21	21	20	20	20	20	20	20
17	23	23	23	22	22	22	22	22	21	21	21	21
16	25	25	24	24	24	24	23	23	23	23	22	22
15	27	27	26	26	25	25	25	25	24	24	24	24
14	29	29	28	27	27	27	27	27	26	26	26	25
13	31	30	30	30	30	29	29	29	28	28	28	27

ASCENSION TABLE

TIME	ALT (FT)
0:10	120
0:20	240
0:30	360
0:42	500
0:50	400
1:02	600
1:10	830
1:17	1000
1:46	1250
2:10	1500
2:34	1750
2:56	2000
3:43	2500
4:31	3000
5:21	3500
6:09	4000
7:00	4500

Table 6-6B. Conversion chart for 30-gram helium (pilot) balloon.

LOAD	FORWARD THROW DISTANCES FOR FIXED-WING AIRCRAFT		
	C-5	**C-130**	**C-17/C-141**
Personnel or Door Bundle	229 M (250 YD)	229 M (250 YD)	229 M (250 YD)
Heavy Equipment	668 M (730 YD)	458 M (500 YD)	668 M (730 YD)
CDS		503 M (550 YD)	686 M (750 YD)
TTB		147 M (160 YD)	147 M (160 YD)
NOTE: To convert yards to meters, multiply yards by 0.9144. To convert meters to yards, divide meters by 0.9144.			

Table 6-7. Forward throw distances for fixed-wing aircraft.

b. **VS-17 Marker Panel, Aerial**. The two-sided VS-17 marker panel (NSN 8345-00-174-6865) measures 2 feet wide by 6 feet long. One side is international orange. The other side of the panel is cerise (red). Six tie-down points permit attachment to stakes. The short ends in the stowage pocket have three snap fasteners. When folded, the panel's olive drab green should show. Pathfinders should display the panel side whose color contrasts best against the surrounding area.

c. **Light, Marker, Ground Obstruction**. One BA-200 battery powers this "beanbag light" (NSN: 6230-00-115-9996). Interchangeable colored plastic domes offer different colors of light. These markers work well in light holes or on the surface. The ground crew secures the markers with tent pegs or by filling the bottom with sand or rocks.

d. **Raised-Angle Marker**. This locally constructed RAM marks the PI on CARP DZs. It consists of five VS-17 panels (Figure 6-5, page 6-15).

e. **Whelen Light**. This light attaches to the top of one of two types of battery originally used with the AN/PRC-77. To place the light into operation, the user seats it on top of the battery. Different colored domes offer different colors of light. The unit buys this light locally. The batteries are as follows:

- Dry battery BA-4386/U.
- Lithium battery BA-5598/U.

f. **M-2 Light Baton**. Two BA-30s power this flashlight (NSN: 6230-00-926-4331). Different lenses (stored in the base compartment) change the color of the light. This light works best in a light hole or on top of the ground attached to a tent peg.

g. **Aerial, Marker, Distress**. This omnidirectional flashing (strobe) light (NSN 6230-00-67-5209) has a very long range. An optional directional cover snaps on top for tactical operations. Other snap-on caps change color and function such as the black cap, which makes the strobe light invisible except to devices that can "see" infrared.

h. **Mirror, Emergency Signaling, Type II**. Pathfinders can use the signal mirror (NSN 6350-00-105-1252) to signal aircraft by reflected sunlight. The back of the mirror has a set of instructions for proper use and aiming. The signal mirror works even on hazy days. It works in all directions--not just when the user faces the sun--and the intended viewer can see it from as far away as the horizon.

i. **Pilot Balloon**. Pathfinders use the PIBAL 10- or 30-gram rubber balloon to measure the mean effective wind. They fill the balloon with helium until the balloon inflates to the specified circumference. National stock numbers for PIBALs follow:

- NSN 6660-00-663-7933, balloon, meteorological, 10-gram.
- NSN 6660-00-663-8159, balloon, meteorological, 30-gram.

j. **Lighting Unit**. This light (NSN 6660-00-839-4927) attaches to the PIBAL for night operations. Overinflating the PIBAL compensates for the weight of the light so it can ascend at the same rate as it would without the light. Water or any other fluid will activate the PIBAL's wet-cell battery. Below 50 degrees Fahrenheit, warm water activates the light faster.

k. **Drift Scale**. This slide-type scale uses a 90-degree angle to measure the ascent of the PIBAL. Pathfinders use the drift scale to compute the mean effective wind. TSC locally produces the drift scale. Pathfinders can also use the *pocket transit* (small enough to carry in a pocket) *theodolite* (NSN 6675-00-861-7939) with built-in *clinometer* (NSN 6675-00-641-5735); or they can use the separate *clinometer* (NSN 6675-01-313-9730).

l. **AN/PRC-119A (SINCGARS) Radio**. This man-portable radio NSN 5820-01-267-9482) allows FM radio contact with aircraft. It also permits NAVAID for aircraft with FM-homing capabilities. Without power-increasing accessories, it transmits between 4 and 16 kilometers.

m. **AN/PRC-113 (Have Quick) Radio**. This man-portable UHF/VHF AM radio (NSN 5820-01-136-1519) has a quick, jam-resistant, ECCM transceiver. Pathfinders use it for short ranges--5 to 16 miles--for tactical, ground-to-ground, or ground-to-air communication.

6-12. COORDINATION

The drop zone coordination checklist provides the DZST leader with a tool for coordinating before the mission without having to communicate with the aircraft (Figure 6-6, page 6-18).

6-13. SUPPORT REQUIREMENTS

The following support requirements apply to multiple aircraft formations (USAF aircraft), personnel, and equipment; or to single aircraft operations on DZs more than 2,100 meters in length:

a. **Control Group**. The DZST leader ensures the DZST sets up on the DZ and is operational at least one hour before the drop. A complete control group consists of the following:

(1) DZST leader.
(2) Assistant DZST leader (DZST qualified).
(3) Malfunction officer or NCO with camera.
(4) Parachute-recovery detail (with saw and tree-climbing equipment).
(5) Radios at least one each for the DZST leader and assistant leader.
(6) Anemometers, one each for the DZST leader and the assistant DZST leader (AN/PMQ-3A or other acceptable anemometer).
(7) Smoke grenades and flares, as needed.
(8) Vehicles, as needed.
(9) Road guards, as needed.
(10) Military police, if needed to control vehicles and spectators.
(11) Two medical personnel with front line ambulance (FLA), except on CDS drops.
(12) Boat detail, if needed.

b. **Rescue Boat**. For a DZ within 1,000 meters of water (from any edge), and for water more than 4 feet deep or 40 feet wide, pathfinders must arrange for at least one boat. Each boat must have a boat operator and an assistant boat operator. The boat's passenger rating must indicate that the boat can safely carry both operators, all parachutists exiting on the first pass, a working engine, plus a life vest for each operator and potential passenger. Ten minutes before TOT, the operators must have the boat in the water and its engine running. They must have a life jacket on board the boat for the first pass. For deliberate water jumps, each parachutist wears a life vest.

NOTE: Jumpers wearing B5s or B7s need no life jackets.

c. **Other Support**. The units may supplement these requirements based on the type of drop, size of the airborne operations, number of aircraft involved, and number of parachutists.

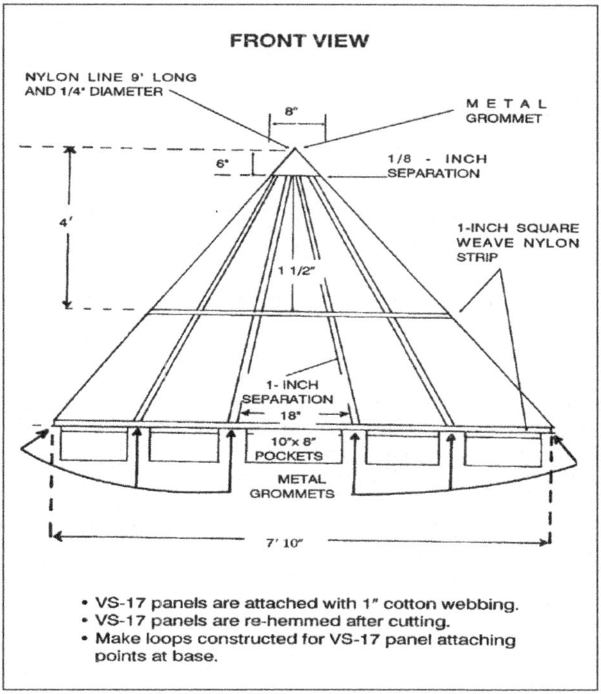

Figure 6-5. Raised-angle marker.

d. **Single Aircraft**. For a single aircraft with less than 20 seconds exit time and when the DZ measures less than 2,100 meters in length, a partial control group with at least the following personnel and equipment can handle the DZ:

(1) The DZST leader.

(2) The assistant DZST leader, who can support this type of DZ without having DZSTL qualification.

(3) Anemometers.

(4) One medic (with FLA). If he has received specialized training in these duties, as discussed in Chapter 20, FM 57-220, the DZST leader can perform this additional duty. CDS drops do not require the presence of a medic.

(5) Malfunction officer or NCO with camera.

(6) Parachute recovery detail.

(7) Radio.

(8) Compass.

(9) Smoke grenades or flares, as needed.

e. **Basic Equipment List**. The DZST leader should maintain an inventory of the following basic equipment to support the mission:

(1) VS-17 panels.

(2) Smoke grenades or flares.

(3) White lights such as an M-2 light baton.

(4) Air traffic control light (B-2).

(5) Signal mirror.

(6) Strobe light.

(7) Binoculars.

(8) Anemometer required for personnel and heavy equipment drops, recommended for measuring the wind before all types of drop.

(9) Compass.

(10) PIBAL kit with helium.

(11) Night vision goggles, for night drops.

(12) Other equipment as needed, based on premission coordination or unit SOP.

6-14. DROP ZONE SUPPORT TEAM LEADER'S DUTIES

The DZST leader establishes and operates the DZ. He selects the locations of the control center, point of impact (PI), and release point. He bears the ultimate responsibility for accomplishing the mission. Specifically, the leader--

a. Makes sure the DZ reaches full operational status one hour before drop time.

b. Conducts premission coordination.

c. Opens the DZ through range control. After the mission, accounts for all personnel, air items, and equipment, then closes the DZ.

d. At least one hour before the drop, reconnoiters the DZ on the ground or from the air for obstacles or safety hazards.

e. Establishes communication with departure airfield control officer NLT one hour before drop time.

f. Controls all ground and air MEDEVACs.

g. Submits postmission reports to the appropriate agencies.

h. Operates all visual acquisition aids.

i. Ensures someone relays no-drop signals to the drop aircraft.

j. Ensures all DZ markings display correctly.

k. Establishes a ten-minute window. Ensures pathfinders continuously monitor surface winds, starting NLT twelve minutes before TOT. This includes the ten-minute window plus two extra minutes to relay a *NO-DROP* signal, if needed. For example, if TOT is 0700 hours, then the ten-minute window (plus two minutes) begins at 0648 hours. If at any time during the ten-minute window the winds exceed allowable limits, the

DZST leader relays a *NO-DROP* to the aircraft. Once he calls a *NO-DROP*, he establishes a new ten-minute window (without an extra two minutes). For example, if the winds pick up at 0655 hours, the leader calls a *NO-DROP*. The new ten-minute window counts from the time of the *NO-DROP* and extends to the new TOT ten minutes after that, at 0705 hours.

l. When the DZ is 2,100 meters in length or longer, when exit time is 20 seconds or more, or for a multiple aircraft operation, takes surface wind readings from the control center location and from the highest point of elevation on the DZ.

m. Calls a no-drop when surface winds exceed the limits shown in Table 6-8, page 6-19.

6-15. CONTROL CENTER

The DZST leader controls and observes the airborne operation from the control center. Pathfinders also take wind readings here. The DZST leader should position all radios, signaling devices, and appropriate forms at the control center. The type of mission determines the location of the control center.

a. **Personnel Drops.** Locate the control center at the PI.

b. **CDS Drops.** Locate the control center 150 yards to the 6 o'clock of the PI.

c. **Free Drops, Heavy Equipment, and AWADS (Ceiling Less Than 600 Feet).** Locate the control center off the DZ where you can see both the approaching aircraft and the PI. For example, the wood line might obstruct the leading edge. If so, it would not make a good control center location for these types of drops.

d. **All GMRS and VIRS DZs.** Locate the control center at the RP.

6-16. SIGNALS

When voice control does not work, the ground support team uses visual signals to the aircraft. Two of the most important visual signals are NO-DROP and MISSION CANCELLATION.

a. To communicate a NO-DROP situation to the aircraft, scramble the shape designator and remove the markings or any other previously coordinated DZ signal.

(1) The drop aircraft pilot should continue to fly racetracks until you give the signal indicating clearance to drop. You can signal NO-DROP when--

- Winds exceed the maximum limitations for that type of drop.
- When you see vehicles moving on the DZ.
- When you see anything else unsafe on the DZ.

(2) Decide in mission coordination how many no-drop passes the pilot must fly before the mission is automatically cancelled and the pilot can begin his return to base.

b. Cover signals for CLEAR TO DROP also. You may decide to indicate clearance to drop by emplacing DZ markings. You can also use this means if you have no smoke. If you plan to use smoke, decide what each color of smoke will mean, but avoid using red to mean CLEAR TO DROP.

c. At night, your CLEAR-TO-DROP signals could include any means coordinated in advance such as shade-designator illumination, a flashing white light, a green light, and so on.

Confirm the following:

TYPE OF LOAD	SURFACE WIND (IN KNOTS)
Personnel (land)	13
Personnel (water)	17
HALO or HAHO	18
Equipment without ground disconnects	13
Equipment with ground disconnects	17
CDS or door bundles using G-13 or G-14 parachutes	20
USAF tactical training bundles	25
High-velocity CDS	No Restrictions
NOTE: For USAF personnel and additional equipment, see Chapter 2, AFI 13-217.	

Table 6-8. Surface wind limits for airdrops.

6-17. DETERMINATION OF RELEASE POINT LOCATION

To determine a release point on a GMRS drop zone, Air Force fixed wing VIRS drop zone, or Army rotary wing VIRS drop zone (Figure 6-7, page 6-20)--

a. **Step 1.** Determine the location of the point of impact, where the first parachutist or load will land. The following criteria should apply:

(1) ***Door Bundles.*** Locate the PI at the leading edge of the tree line, centerline.

(2) ***Personnel.*** Locate the PI on the leading edge of the 100-meter buffer zone, centerline.

(3) ***CDS and Heavy Equipment.*** See AF Form 3823 for the surveyed PI location.

(4) ***CDS and Heavy Equipment on Nonsurveyed DZs.*** See the standard PI locations for tactical assessments.

b. **Step 2**. Determine the drift direction and distance.

(1) Compute a D = KAV formula for your mission.

(2) Determine back azimuth of wind direction.

(3) From the PI location, pace off the distance computed from the D = KAV formula into the wind.

(4) If the wind blows to an azimuth of 250 degrees, then use 070 degrees for the pace-off azimuth.

c. **Step 3**.

(1) Determine the forward throw.

(2) Once you have paced off the drift, walk off the forward throw.

(3) To do this, shoot a back azimuth of the drop heading and walk the distance of the forward throw (for Air Force aircraft forward throw, refer to Table 6-7, page 6-12).

(4) For rotary wing aircraft, divide the drop airspeed in half, and express in meters. For example, the drop speed of 90 knots equals 45 meters of forward throw. The release point is at the end of the forward throw (Table 6-7, page 6-12).

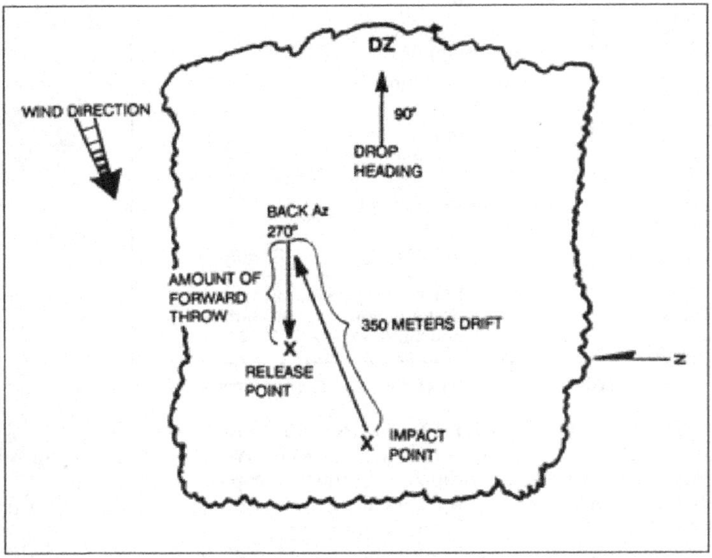

Figure 6-7. Release point location.

6-18. GROUND-MARKING RELEASE SYSTEM

The GMRS offers the DZST a way to identify the release point to the drop aircraft without using a radio. The pilot uses the ground markings to adjust his flight path 100 meters to the right of the corner panel or light, and parallel to the approach-corner panel or light axis. (This discussion uses the words "panel" and "light" interchangeably.)

a. **Patterns**. Use VS-17 panels to mark the DZ with an inverted "L," "H," or "T" pattern.

(1) ***Inverted "L" Pattern.*** The inverted "L" has four panels:

- The approach panel.
- The corner panel.
- The alignment panel.
- The flanker panel.

(2) ***"H" and "T" Patterns***. Align these other panels with and orient them on the corner panel. Due to side-angle-vision limitations in the C-5 and C-141, use the seven-panel "H" and six-panel "T" patterns for these aircraft.

b. **VS-17 Panels**. Figure 6-8, page 6-22, shows panel emplacement for "H" and "T" patterns.

(1) ***Corner Panel***. Set up the corner panel 100 meters to the left of the RP (as seen from drop heading).

(2) ***Alignment Panel***. Place the alignment panel 50 meters to the left of the corner panel.

(3) ***Approach Panel***. Place the approach panel 50 meters in front of the corner panel, as seen from the drop heading.

(4) ***Flanker Panel***. Place the flanker panel 150 meters to the left of the alignment panel, as seen from the drop heading.

c. **Drop Zone Support Team .** Place the markings where obstacles will not mask the pilot's line of sight. As a guide, use a mask-clearance ratio of 1 to 15 units of horizontal clearance (Figure 6-9, page 6-23).

(1) For example, suppose you must position a DZ marker near a terrain mask, such as the edge of a forest on the DZ approach. The trees measure 10 meters (33 feet) high. The markings would require 150 meters (492 feet) of horizontal clearance from the trees (Figure 6-10, page 6-24).

(2) If any of the GMRS markings fall within a 15-to-1 mask clearance ratio on the approach end of the drop zone, you can place an Army code letter (H, E, A, T) or a far (marker) panel on the trailing edge of a DZ. However, before doing so, you must have coordinated this during the DZST briefing or during an aircrew mission briefing. Using a code letter will distinguish this DZ from other DZs in the area. The code letter is located at the end of the DZ or where the pilot can see it best.

d. **Lights**. At night, replace panels with lights--use one light for each panel. Use directional lights for the approach, corner, alignment, and flanker. If necessary, you can use the directional light holes for the far code letter and line up the base light with the corner light. Figure 6-10 (page 6-24) provides construction requirements for Army code letters. Mark the release point with some type of identifiable light source to distinguish it from all other DZ markings.

6-19. ARMY VERBALLY INITIATED RELEASE SYSTEM

The Army VIRS method establishes the release point on the DZ via radio communications. If tactically feasible, a code letter can mark the RP. However, the aircraft initiates the drop on verbal command from the ground.

a. Emplace a standard Army code letter with VS-17 panels (for daytime operations) at the release point. Position the base panel of this code letter exactly on the RP. Use code letter H, E, A, or T. Make the letter at least two panels high by one panel wide (Figure 6-10, page 6-24).

b. Emplace a flank panel to the left of a code letter at a distance of 200 meters or on the edge of the DZ, whichever is closer. Emplace a far panel 500 meters from the code letter along the drop heading or at the end of the DZ, whichever is closer. Position each panel with its long axis parallel to the drop heading and raised 45 degrees back toward the code letter. The GTA radio operator positions himself at the release point.

c. At night, replace the panels with lights. Make the code letter at least four lights high by three lights wide, and leave 5 meters between each light. To limit ground observation, you may place the code letter, flank, and far light in holes.

- Place the code letter and far light in directional holes.
- Place the flank light in a bidirectional hole.

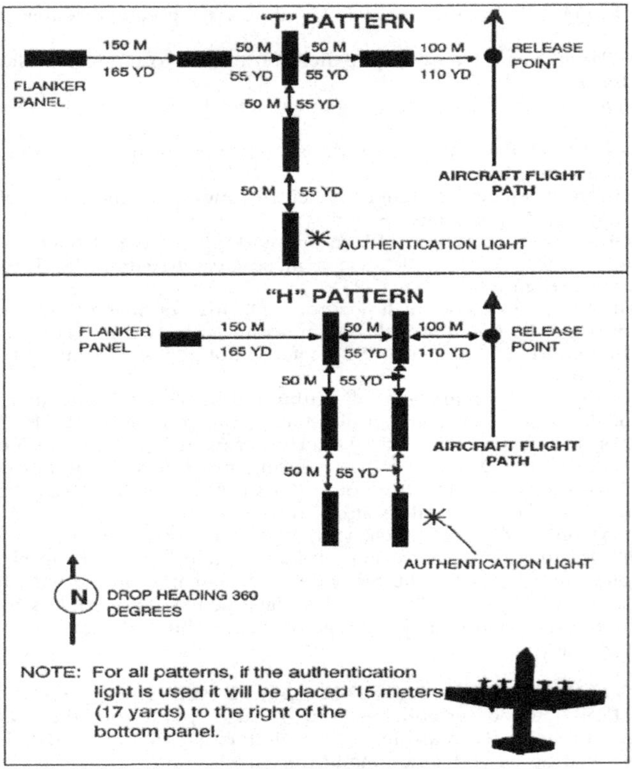

Figure 6-8. Panel emplacement.

d. If the RP falls off the DZ, hiding the markings, or if the GTA operator cannot see the aircraft, change the parachute drop to a jumpmaster-directed release operation using the wind streamer vector count (Figure 6-11, page 6-25).

e. If the RP falls off the DZ and the jumpmaster cannot count the wind streamer vectors, he may release early or late. To release late, he determines how many seconds the aircraft must fly past the GTA operator before he begins the release. For example, multiplying a drop speed of 70 knots by the conversion factor of 0.51 equals 36 meters per second of flight (knots x 0.51 = meters per second).

f. On an Army VIRS DZ, where trees obscure the RP, you may use an offset (Figure 6-12, page 6-26).

(1) An UH-60A is traveling at a speed of 70 knots (about 36 MPS).

(2) RP location is 108 meters into the wood line.

(3) As the aircraft passes over the code letter, the GTA operator begins counting, "ONE THOUSAND, TWO THOUSAND, THREE THOUSAND..."

(4) At the end of three seconds, the GTA operator transmits, "EXECUTE, EXECUTE, EXECUTE."

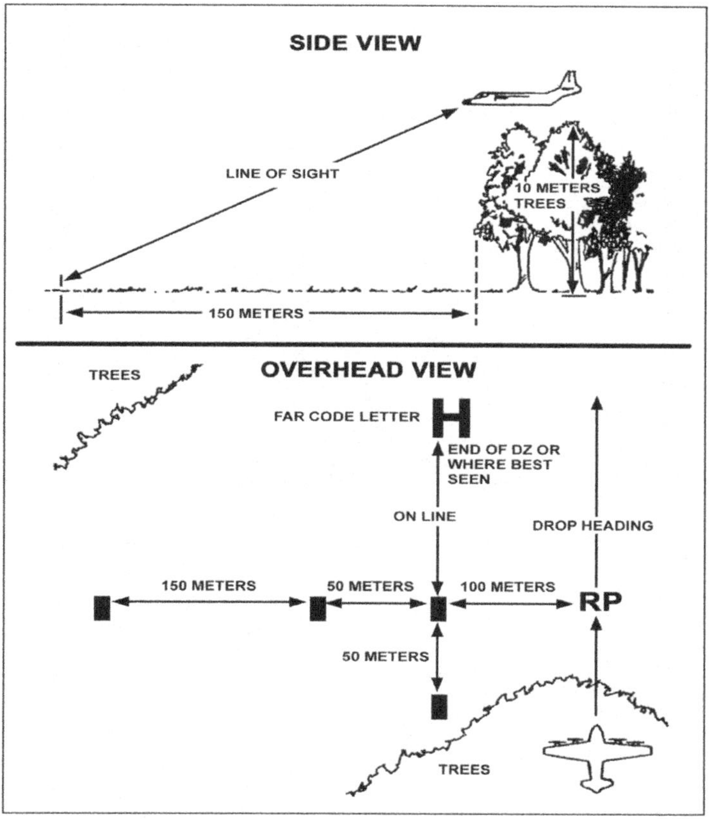

Figure 6-9. The 15-to-1 mask clearance ratio.

6-20. AIR FORCE VERBALLY INITIATED RELEASE SYSTEM

When using USAF VIRS, the DZST leader verbally guides the pilot over the RP to align the aircraft for the drop. Figure 6-13, page 6-27 shows an example USAF VIRS transmission. The leader-

a. Uses this method when the tactical situation prevents use of regular markings or when aircraft pilots could not see regular markings from the air.

b. Chooses the VIRS DZ just as he would choose a GMRS or rotary-wing VIRS release point. (Figure 6-7, page 6-20, shows how to determine the location of the release point.) Unlike those release points, however, the USAF VIRS DZ RP requires no markings.

c. Sets up communications with the drop aircraft and at least two FM, VHF, and UHF radios on the DZ.

d. Transmits concise instructions to the aircraft. To align the aircraft on the desired inbound heading, he gives left and right turns. When the aircraft lines up on course, the pathfinder signals STOP TURN. About five seconds before the release, or at some other moment (as briefed), he signals STANDBY.

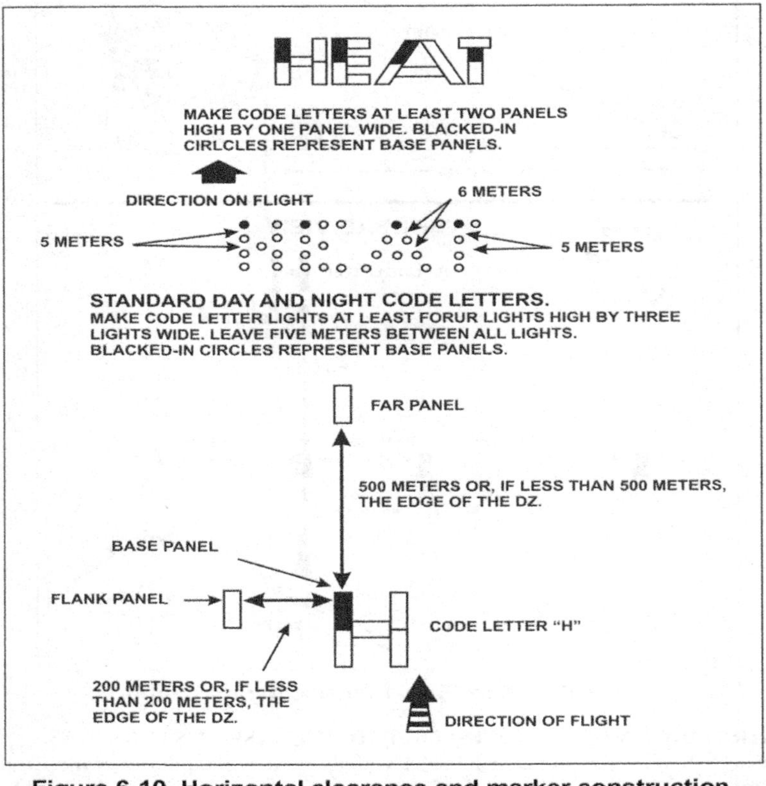

Figure 6-10. Horizontal clearance and marker construction.

e. When the aircraft reaches the predetermined release point, gives an EXECUTE three times.

f. When transmitting the MEW to the aircraft, make sure to identify it as such. State the altitude used to obtain it. Also provide pertinent details about any erratic winds or wind shears reported by other aircraft.

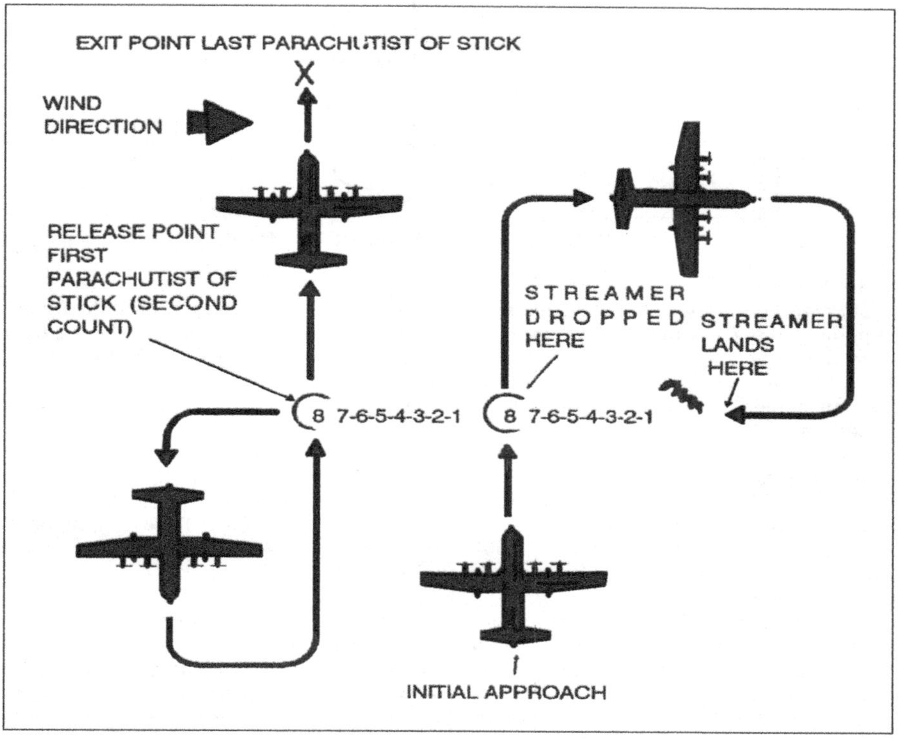

Figure 6-11. Wind streamer vector count.

6-21. AIR FORCE COMPUTED AIR RELEASE POINT

The DZST leader and DZ party mark the point of impact on a surveyed DZ. The aircraft navigator computes the release point from the air.

a. **Coordination**. Coordinate authentication markings with the aircrew.

b. **Code Letter Elevation**. Never elevate the panels in the code letter.

c. **Personnel Buffer Zone**. If the PI is located closer than 300 yards (274 meters) from any edge of the drop zone, allow a 200-yard (183-meter) buffer zone for personnel on the trail edge of the DZ.

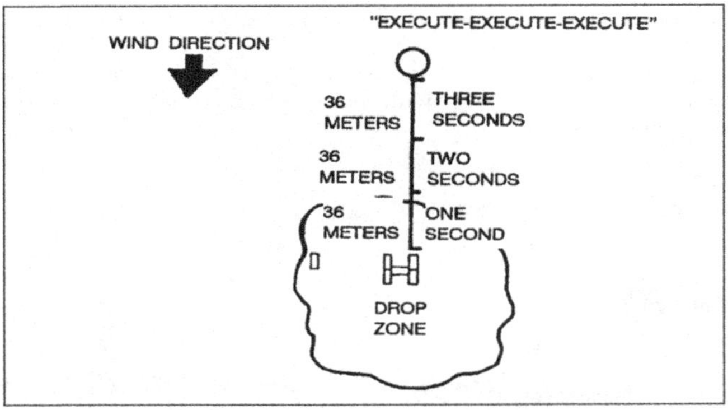

Figure 6-12. Example Army VIRS offset.

d. **Markers**.

(1) ***Day Operations***. For day operations (Figure 6-14, page 6-28), pathfinders mark the point of impact with a RAM (raised-angle marker, Figure 6-5, page 6-15) and an Air Force code letter, which could be a "J," "C," "A," "R," or "S" (Figure 6-15, page 6-29). This applies to rectangular drop zones. They use "H" or "O" for circular drop zones. Each letter must measure at least 35-feet square, and requires at least nine panels flat on the ground.

(2) ***Night Operations***. For night operations (Figure 6-16, page 6-30), mark the point of impact with--

(a) A code letter (made of lights) placed at the point of impact.

(b) One flanker light placed 250 meters to the left and right of the shape designator.

(c) A trailing edge light placed 1,000 meters from the shape designator or at the trailing edge of the surveyed DZ, whichever is closer to the PI.

e. **Authentication**.

(1) ***Day Operations.*** During day operations, authenticate or identify different sites by specifying drop times, drop headings, or alternating panel colors.

(2) ***Night Operations.*** During night operations, authenticate or identify different sites by replacing one light in the code letter with any color light except white.

f. **Circular Markings.** To determine the desired drop heading, coordinate circular DZ markings 24 hours in advance.

g. **Smoke.** You may display smoke, in any color besides red, next to and downwind of the RAM or code letter. This helps show the wind direction, so it also helps the pilot visually acquire the DZ.

h. **Point of Impact.** The Army ground unit commander may position the PI anywhere on the "surveyed DZ" and, as needed, insert his forces to accomplish the mission. He must inform the supporting airlift and the DZST leader of the PI location early enough for them to plan the mission. If the GUC chooses a PI closer than 300 yards (274 meters) from any edge of the surveyed DZ, then he accepts the responsibility for an off-the-DZ impact.

INITIAL RADIO COMMUNICATION		
	AIRCRAFT:	*T2S26, THIS IS HERC 30, OVER.*
	DZST:	HERC 30, THIS IS T2S26, AUTHENTICATE DELTA FOXTROT, OVER.
	AIRCRAFT:	*T2S26, THIS IS HERC 30, GOLF, OVER.*
	DZST:	HERC 30, T2S26--NOT IN SIGHT. CONTINUE, OVER.
ONCE THE AIRCRAFT IS ABOUT ONE MINUTE OUT		
	DZST:	HERC 30, THIS IS T2S26, HAVE YOU IN SIGHT, TURN LEFT.
	DZST:	(ELIMINATE CALL SIGNS) TURN LEFT. . . STOP TURN.
	DZST:	ON COURSE, STAND BY. (5 SECONDS FROM DROP).
	DZST:	EXECUTE, EXECUTE, EXECUTE, T2S26 OUT.

Figure 6-13. Example USAF VIRS transmission.

6-22. ASSAULT ZONE AVAILABILITY REPORT

The Air Force lists all available approved drop zones in its AZAR. This list stems from input provided by 21 AF, McGuire AFB, New Jersey, and from 22 AF, Travis AFB, California. The AZAR identifies CONUS drop zones, landing zones, and extraction zones available for use by the Air Mobility Command.

6-23. DROP ZONE SURVEY

Air Force Form 3823 (front and back, Figure 6-17A and Figure 6-17B, pages 6-32 and 6-33) includes all needed information about the drop zone. The boxes on the form are self-explanatory. If needed, Air Force Instruction (AFI) 13-217 provides additional detailed instructions. Though AF Form 3823 replaces MAC Form 339 (Figures 6-18A and 18B, pages 6-34 and 6-35), pathfinders who have already documented a DZ on the old form can keep using that one until the DZ changes or until no one uses that DZ anymore.

6-24. TACTICAL ASSESSMENT

During contingency, wartime, and major training exercise participation, DZST leaders may have to tactically locate and assess a potential drop zone for follow-on airdrop resupply or reinforcement missions.

a. Normally, the Air Force special tactics teams (STTs) use the AF Form 3823 for this reconnaissance type mission. However, in the absence of a special tactics team (STT), the DZST leader can use the checklist shown in Figure 6-19, page 6-36 to tactically assess the DZ.

b. Tactically assessed DZs may *only* receive airdrops if--

(1) During training events, the airdrop is located within a military reservation or on property the US government has leased.

(2) The supported service accepts responsibility for any damage that occurs as a result of the airdrop activity.

(3) All parties allow adequate time for safe and effective planning.

c. After completing the tactical assessment and receiving approval from the Air Force, those conducting the airdrop follow guidelines for CARP markings (Table 6-9, page 6-36).

(1) ***CDS Drops***.

(a) *C-130*. For C-130 CDS drops during the day, choose a PI at least the following distance from the leading edge of the DZ and centerline:

- Daytime--200 yards (183 meters).
- Nighttime--250 yards (229 meters).

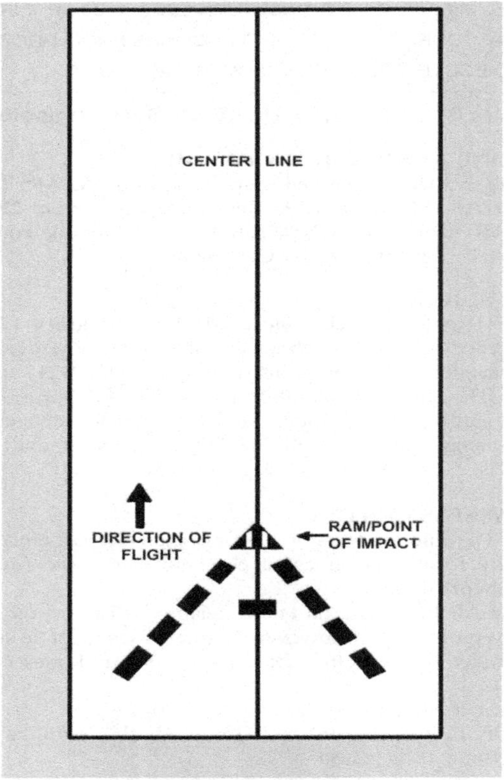

Figure 6-14. Drop zone placement (day).

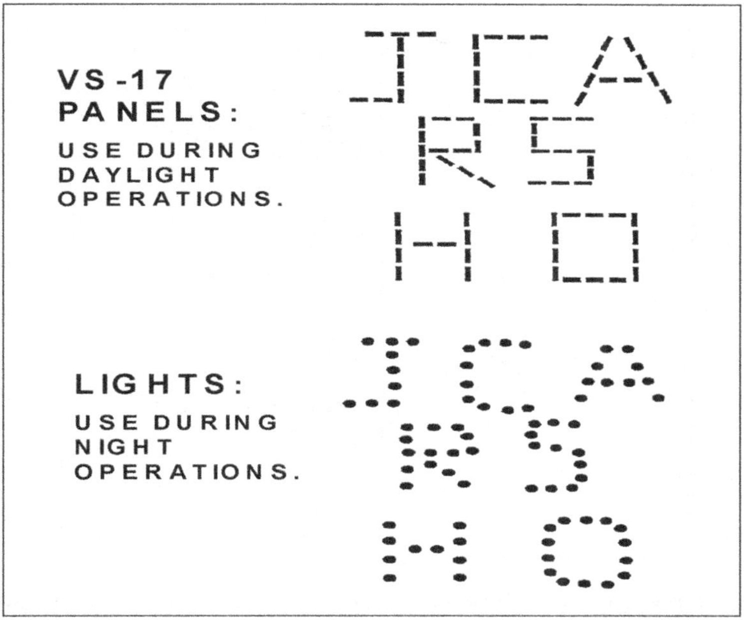

Figure 6-15. Code letters.

(b) *C-141/C-17*. For C-141 or C-17 CDS drops, choose a PI at least the following distances from the leading edge of the DZ:

- Daytime--225 yards (206 meters).
- Nighttime--275 yards (251 meters).

(2) ***Personnel Drops***. For all personnel drops by USAF fixed-wing aircraft, choose a PI at least the following distances from the leading edge of the DZ:

- Daytime--at least 300 yards (274 meters).
- Nighttime--at least 350 yards (320 meters).

(3) ***Heavy Equipment Drops***. For all heavy equipment drops by USAF fixed-wing aircraft, choose a PI at least the following distances from the leading edge of the DZ:

- Daytime--at least 500 yards (457 meters).
- Nighttime--at least 550 yards (503 meters).

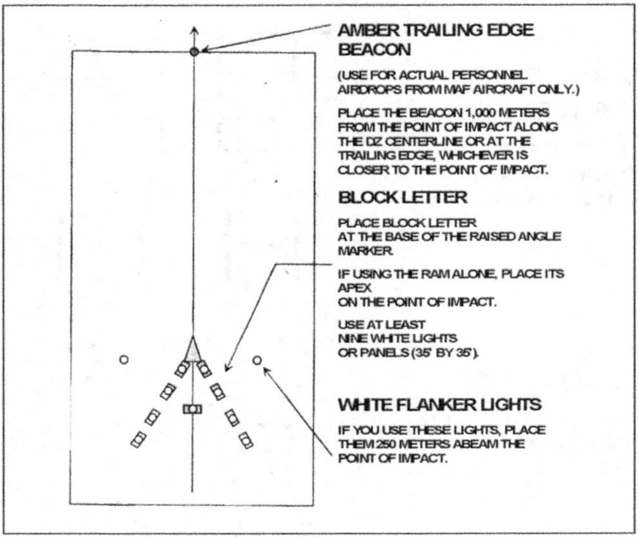

Figure 6-16. Drop zone placement (night).

6-25. CONTROL LOG FOR AIRDROP, AIRLAND, OR EXTRACTION ZONE

The AMC Form 168 serves as a scorecard for the Air Force. Since the aircrew on the CARP drop zone computes the release point, the Air Force must document the crew's performance. The DZST leader completes the blocks on AMC Form 168 (Figure 6-20, page 6-37). In Blocks 2, 3, and 4, he can name the same or different people. After completing the form, he forwards it to his air operations officer, who submits it through the chain of command to the Air Force representative.

a. ***LOCATION***. Write the name and grid coordinates for the DZ, LZ, and EZ.

b. ***SST AND UNIT***. Write the names of the members of the special tactics team and of the unit controlling the DZ.

c. ***DZ–LZ–EZ CONTROL OFFICER(S) AND UNIT(S)***. Write the name of the controlling individual(s) and unit(s).

d. ***DROP ZONE SAFETY OFFICER AND UNIT***. Write the name of DZ safety officer and unit.

e. ***LEGEND***. This section of the form explains both the abbreviations printed on this form and the abbreviations you can use when completing it.

f. ***LINE NO***. On any given DZ mission, each line number represents an individual pass over the DZ, even on a multiple-aircraft DZ. For example, use three lines for a three-ship operation. Use a line and line number for each no-drop pass, too. Document the reason for each no-drop pass in the REMARKS block.

g. ***TYPE ACFT***. Write the type aircraft, for example, "C-130."

h. ***UNIT***. Write the name of the flight crew's unit. (This is usually numeric.)

i. ***CALL SIGN***. If you have radio communications with the pilot, write the correct call sign.

j. ***PILOT – NAVIGATOR***. Write the last name and rank of both the pilot and the navigator.

k. ***TYPE MSN***. Based on your legend, write the letter that you are using to designate the mission type.

l. ***ETA***. Write what time you estimate the aircraft will arrive over the target. You can use a hard time or a block time.

m. ***ATA.*** Write the actual time of arrival.

n. ***ATD***. Write the actual time of departure.

o. ***STRIKE RPRT***. The information in these boxes (yards and clock direction) is the actual purpose of the AMC Form 168. Observe the first parachutist, bundle, or heavy equipment platform from the control center. At night, use NVG. If you maintain radio communication with the drop aircraft, relay the strike reports to it so the aircrew can adjust on following passes over the DZ.

(1) ***YDS***. Determine the distance to the first parachute in yards. If the first parachute lands within 25 yards of the point of impact, write the letters "PI" in this block to show that the parachute made a direct hit.

(2) ***CLOCK***. Determine the clock direction relative to the point of impact (12 o'clock is the magnetic drop heading).

p. ***AL EX***. For an airland or extraction. If you could not maintain visual contact with the first parachute, especially during multiple aircraft operations, then use the following method to determine whether to score parachutists satisfactory or unsatisfactory ("S" or "U"):

(1) ***S***. If 90 percent of the parachutes land on the DZ, write an "S" in this column.

(2) ***U***. If less than 90 percent land on the DZ, write a "U" in this column.

q. ***SURF WIND***. Write down the highest wind reading taken during the 10-minute window for that pass.

r. ***SCORE METHOD***. Write the first letter of the method you used to determine the distance from the PI to the first parachute:

- Write an "E" if you estimated the distance.
- Write a "P" if you paced off the distance.
- Write an "M" if you measured the distance.

s. ***MEAN EFFECTIVE WIND***. If you use a pilot balloon on the DZ, determine a MEW and note it in this block. (MEW is the average wind speed between the ground and the drop altitude.) If you have established radio communications with the aircraft, give the aircrew the MEW before they make their first pass. This will help them calculate their release point.

(1) ***TIME***. Note the time that you determined the MEW.

(2) ***ALT***. Note the altitude you used to determine the MEW.

(3) ***DIRECTION and VELOCITY***. Measure the magnetic azimuth to the pilot balloon, and write down the reciprocal heading (MEW direction). When you report the MEW, give the altitude that you used to compute it. If you noted any erratic winds or wind shears, note that here also.

t. ***REMARKS (Continue on Reverse)***. Write down anything about the DZ operation that you think the Air Force might like to know before debriefing the pilot.

EXAMPLE

AIRBORNE UNIT ASSUMES RESPONSIBILITY FOR PERSONNEL INJURY AND EQUIPMENT DAMAGE ON DZ		
DROP ZONE SURVEY	1. DZ NAME	2. LOCATION
	3. MAP SERIES/SHEET NUMBER/EDITION/DATE OF MAP	

4. SURVEY APPROVAL / DISAPPROVAL DATA			
4A1. DATE SURVEYED	4A2. NAME AND GRADE OF SURVEYOR	4A3. PHONE NUMBER (DSN)	4A4. UNIT

4B. DROP ZONE APPROVAL/DISAPPROVAL A = APPROVED D = DISAPPROVED	FOR	CDS/CRS	PER	HE	MFF	SATB	CRRC	HSLADS	HVCDS	
	DAY									
	NIGHT									

4C. DATE APPROVED FOR GROUND OPERATIONS	NAME AND GRADE OF APPROVING AUTHORITY	PHONE NUMBER (DSN)	SIGNATURE
	UNIT AND LOCATION		
4D. DATE OF SAFETY OF FLIGHT REVIEW APPROVED	NAME AND GRADE OF APPROVING AUTHORITY	PHONE NUMBER (DSN)	SIGNATURE
	UNIT AND LOCATION		
4E. DATE OF MAJCOM APPROVAL	NAME AND GRADE OF APPROVING AUTHORITY	PHONE NUMBER (DSN)	SIGNATURE
	UNIT AND LOCATION		

5. COORDINATING ACTIVITIES		
A. DZ CONTROLLING AGENCY OR UNIT	B. MEMORANDUM OF UNDERSTANDING/LAND USE AGREEMENT YES ☐ NO ☐ ATTACHED ☐	C. PHONE NUMBER (DSN)
D. RANGE CONTROL		E. PHONE NUMBER (DSN)

6. DZ DIMENSIONS (YDS/MTRS) (FOR CIRCULAR DZ, ENTER RADIUS ONLY)					
A. LENGTH	B. WIDTH	C. RADIUS	TIMING POINT DISTANCES	D. T/P FROM LEADING EDGE	E. T/P FROM DZ CENTERLINE
POINT OF IMPACT DISTANCES FROM DZ LEADING EDGE	F. CDS PI		G. PE PI	H. HE PI	

7. DZ AXIS DATA (OPTIONAL FOR CIRCULAR DZ)			
A. MAGNETIC	B. GRID (UTM)	C. TRUE	D. DATE OF VARIATION DATA

8. GROUND POINT ELEVATION	A. CDS PI	B. HE PI	C. PE PI	D. HIGHEST

9. DZ COORDINATES				
A. SPHEROID	B. DATUM	C. GRID ZONE	D. EASTING	E. NORTHING
F. GPS DERIVED COORDINATES YES ☐ NO ☐	G. POINT OF ORIGIN			

H. POINT	UTM COORDINATES	WGS84 LATITUDE (D-M.MM)	WGS84 LONGITUDE (D-M.MM)
DZ CENTERPOINT			
CDS PI			
PE PI			
HE PI			

I. DZ CORNERS UTM COORDINATES	
LEFT LEADING EDGE	RIGHT LEADING EDGE
LEFT TRAILING EDGE	RIGHT TRAILING EDGE
LEFT TIMING POINT	RIGHT TIMING POINT

AF FORM 3823, FEB 94 (EF-V1)

Figure 6-17A. Example completed AF Form 3823 (new form, front).

DZ NAME

10 DZ DIAGRAM / REMARKS

EXAMPLE

11 PHOTOGRAPH AVAILABLE
YES ☐ NO ☐

12. LOW LEVEL ROUTES
☐ NONE AVAILABLE
☐ ROUTE NAME/DESIGNATOR

AF FORM 3823, (REVERSE) (EF-V1)

Figure 6-17B. Example completed AF Form 3823 (new form, back).

DROP ZONE SURVEY

Example DZ	Ft. Hunter-Liggett, CA
V 7955/Hunter-Liggett Special/1-DMATC/1971	

4\. SURVEY APPROVAL/DISAPPROVAL DATA

23 Dec 88	Mark a. Goodjob, Capt	544-1212	OL-N 1702 MOBSS
15 Jan 89	John B. Closelook, Capt	511-1313	(signature)
	63 MAW/DOXT Norton AFB, CA 92409-5000		
22 Feb 89	Steve E. Okaysum, Maj	555-1414	(signature)
	22 AF/DOXT Travis AFB, CA 94535-5002		

A. APPROVED	DAY	A	A	A	A	A	B	A		
B. DISAPPROVED	NIGHT	A	A	A	A	A	B	A		

5\. COORDINATING ACTIVITIES

Ft. Hunter-Liggett Range Coordinator	533-1515
Ft. Hunter-Liggett Range Control, Call Sign "Hunter 33", 41.05 FM, 229.5/126.2 AM	533-1616

6\. DZ DIMENSIONS (Yds/Mtrs) (For Circular DZ, Enter Radius Only)

3230 yds	800 yds	n/a		700 yds	350 yds (each)
POINT OF IMPACT (PI) DISTANCES FROM DZ LEADING EDGE		550 yds		350 yds	550 yds

7\. DZ AXIS DATA

274	288	289	1989

8. GROUND POINT ELEVATION				
	1220	1220	1225	1378

9\. DZ COORDINATES

Clarke 1866	10	6	39

105 FQ 5540 7860, San Miguelito Ranch House Ruins, 1350 mtrs at 075 degrees Mag to Hvy/CDS PI

F. POINT	UTM COORDINATES	LATITUDE	LONGITUDE
DZ CENTERPOINT	5582 7892	N 35/56/38.6	W 121/16/20.7
	5675 7862	N 35/56/28.3	W 121/15/44.0
	5692 7856	N 35/56/26.3	W 121/15/37.1
	5675 7862	N 35/56/28.3	W 121/15/44.0

5712 7812	5734 7881
5431 7903	5453 7973
5774 7796	5794 7857

10\. ZONE MARKER (ZM) POSITION DATA (Enter Additional ZM Points in Remarks)

HVY PI	200 yds Prior	100 yds Left

MAC Form 339, FEB 89 PREVIOUS EDITION IS OBSOLETE

Figure 6-18A. Example completed MAC Form 339 (old form, front).

11. DZ DIAGRAM / 12. REMARKS

DZ NAME
Example DZ

1. DZ is located within R-2513. Contact Range Control for clearance into the restricted area or the CCT for entry.
2. Racetracks, if used, will be to the North.
3. Obstacles (From DZ center):
   a. Hill, 1890' MSL, 2.3 nm, 077° True.
   b. Hill, 2906' MSL, 1.8 nm, 243° True.
   c. Hill, 2188' MSL, 3.4 nm, 258° True.
4. Aircrews will use extreme caution for high terrain to the West and North of the DZ.
5. Users are requested to coordinate use of the DZ at least three months prior to use due to heavy scheduling.
6. Random PI placement area is 50 yds either side of centerline from the PERS PI to 1000 yds past the HVY/CDS PI.
7. Hilly terrain surrounds the DZ within 5 nm. Winds are normally from the SE. Winds shears may occur if winds are from the NW.

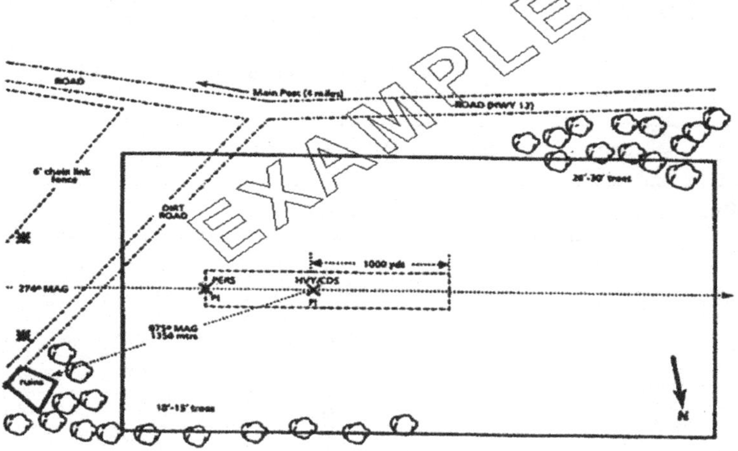

13. PHOTOGRAPHY AVAILABLE

☐ YES ☒ NO

LOW LEVEL ROUTES

☒ NONE AVAILABLE

☐ ROUTE NAME/DESIGNATOR

Figure 6-18B. Example completed MAC Form 339 (old form, back).

1. **DZ name or intended call sign.**
2. **Topographical map series and sheet number.**
3. **Recommended approach axis magnetic course.**
4. **Point of impact location** (Eight-digit grid).
5. **Leading edge centerline coordinates** (Eight digits).
6. **DZ size in meters or yards.**
7. **Air traffic restrictions or hazards.**
8. **Name of surveyor and unit assigned.**
9. **Recommendation for approval or disapproval.** (If disapproval, cite the reason.)
10. **Remarks.** (Include a recommendation for airdrop option, CARP, GMRS, VIRS, or blind drop.)

Figure 6-19. Tactical DZ assessment checklist.

TYPE DROP	AIRCRAFT	DAY or NIGHT	*POINT OF IMPACT
CDS	C-130	Day	At least 200 yards/183m
		Night	At least 250 yards/229m
	C-141/C-17	Day	At least 225 yards/206m
		Night	At least 275 yards/251m
Personnel	All USAF fixed-wing aircraft	Day	At least 300 yards/274m
		Night	At least 350 yards/320m
Heavy Equipment		Day	At least 500 yards/457m
		Night	At least 550 yards/503m

* Point of impact shown in distance from the leading edge of the drop zone and centerline.

Table 6-9. Favorable conditions for airdrops on tactically assessed DZs.

AIRDROP / AIRLAND / EXTRACTION ZONE CONTROL LOG

DATE: 8 SEPT 92

LOCATION	CCT AND UNIT	DZ/LZ/EZ CONTROL OFFICER AND UNIT	DROP ZONE SAFETY OFFICER AND UNIT
HOPSON DZ GL 04592143	SSG BENNET HHC 1/507TH PIR	SSG GUENTHER HHC 1/507TH PIR	SFC PROCTOR HHC 1/507TH PIR

LEGEND

AH–Airland (Heavy)
AL–Airland
CD–CDS
ED–Extraction (Drogue)

EX–Extraction
GM–GMRS
HE–Heavy Equipment
HO–HALO

IL–Inverted "L"
LS–Instrument Landing System
PE–Personnel
RD–Radar Beacon Drop

TC–TT B CDS
TH–TT B Heavy
TP–TT B Personnel
BD–AWADS

SCORE METHOD
E – Estimated
P – Paced
M – Measured

LINE NO	TYPE ACFT	UNIT	CALL SIGN	PILOT-NAVIGATOR	TYPE MSN	ETA	ATA / ATO	STRIKE RPRT YDS	STRIKE RPRT CLOCK	AL EX S	AL EX U	SURF WIND	SCORE METHOD	MEAN EFFECTIVE WIND TIME	MEAN EFFECTIVE WIND ALT	MEAN EFFECTIVE WIND DIR & VEL	REMARKS (Continue on Reverse)
1	C-130	37	38 NOVA	MAJ RAMGERS LT. VICTORIA	HE	1100	1105	PI	PI			5KTS 225	P	1000	1100	255/17KTS	
2					PE	"	1117	150	6			7KTS 230	E	"	"	"	
3					PE	"	1129	200	8			8KTS 234	E	"	"	"	
4					PE	"	1141	50	12			6KTS 230	E	"	"	"	LAST 3 JUMPERS IN TREES
5																	
6																	
7																	
8																	
9																	
10																	
11																	
12																	
13																	
14																	
15																	

MAC FORM FEB 74 168 PREVIOUS EDITION IS OBSOLETE

EXAMPLE

Figure 6-20. Example completed AMC Form 168.

APPENDIX A
OPERATIONAL FORMATS

Pathfinder leaders can use the operational formats described in this appendix to plan and carry out pathfinder operations. These formats are intended as guides and should be modified as required.

A-1. OPERATION PLANNING FORMAT

The pathfinder leader uses the operation-planning format shown in Figure A-1, page A-2, to organize an operation. This format consolidates information about each person or element. The leader can refer to this information during his planning and briefing. This format provides columns for writing in--

a. **ACFT NO**. Write in either the chalk number or the last three digits of the number painted on the tail of the aircraft that will transport the pathfinder.

b. **NAME**. Write in the name of each pathfinder.

c. **LOAD TIME**. Write in what time the pathfinder must be on the aircraft with all of his equipment.

d. **TO TIME**. Write in what time the aircraft will depart the staging area (the takeoff time).

e. **DUTY AND LOCATION**. Write in each pathfinder's mission assignment and location within the operational area.

f. **CALL SIGN AND FREQUENCY**. Write in the radio call sign and frequency for each person operating a radio.

g. **EQUIPMENT**. List all equipment, other than individual equipment, that each pathfinder element will carry.

h. **REMARKS**. Write in any other pertinent information.

A-2. LANDING ZONE AND DROP ZONE CONTROL RECORDS

The leader can make up an LZ/DZ Control Record based on the example format shown in Figure A-2, page A-3 to keep a record of aircraft arrivals, departures, and load types. This record provides information for both ground and aviation commanders. It helps account for personnel and equipment. It can also help leaders initiate or conduct search-and-rescue operations for overdue or downed aircraft. The pathfinder internal radio net(work) operator, located at the control center, normally maintains this record.

a. **Format Headings**.

(1) ***PATHFINDER UNIT***. Enter the pathfinder unit's code or number designation.

(2) ***SUPPORTED UNIT***. Enter the name of the main ground or aviation unit.

(3) ***PERIOD***. Enter the date and time the operation will start and finish. Enter 0001 for the start time of a succeeding day. Enter 2400 for the end time if the operation will continue the next day.

(4) ***OPERATION (AFLD LZ, DZ)***. Enter the name or number of the operation. Cross out items that do not apply. Add any special designation used.

(5) ***RECORDER***. Enter the name of the person who records data on this form.

CLASSIFICATION

PATHFINDER OPERATION PLANNING FORMAT

ACFT NO	NAME	LOAD TIME	TO TIME	DUTY AND LOCATION	CALL SIGN AND FREQUENCY	EQUIPMENT	REMARKS
NA	HEITER	NA	NA	SECTION CO. STAGE FLD DEKKAR STRIP, GL 934730	11B85.34:50	2-PRC-112 RADIOS 1-DWYER METER 1-AN/PMQ-3A ANEMOMETER 2-PRC-113 RADIOS 3-PRC/ 77 RADIOS 1-AN / PRC 41A 1-AN / TRN 30 (EX 1) 24-VS / 17 PANELS 35MX / 290 LANTERNS 3 SETS OF BATONS 1-SE / 11 LIGHT GUN 3 COLORED JACKETS EXTRA SMOKE	BE PREPARED FOR MIXED TRAFFIC AND NIGHT OPERATIONS UNTIL RELIEVED HOMING BEACON FREQ 1750KC.
NA	MARSONNER	NA	NA	GTA RATELO, STAGE FLD	11B85R..34:50		
NA	COFFIN	NA	NA	INTERNAL NET RATELO, STAGE FLD	11B39.66:20		
NA	CASCALASIRA	NA	NA	SIGNALMAN, STAGE FLD	18B67.66:20		
NA	LEACH	NA	NA	SIGNALMAN, STAGE FLD	11B29.66:20		
750	PRUITT	0540	0545	SITE CO. LZ ALBANY GL 055713	J7G38 40:20	1-PRC 41 RADIO 1-PRC 41A RADIO 2-PRC/77 RADIOS 12 MX/290 LANTERNS 1-SE/11 LIGHT GUN 1 SET OF BATONS 6-VS/17 PANELS 8 SMOKE GRENADES	ACCOMPANY B 188TH ON MOVEMENT TO CONTACT OPERATION UNTIL COMPLETED
750	HODGES	0540	0545	GTA RATELO, LZ ALBANY	J7G38R.40:20		
750	CHURCHY	0540	0545	PATHFINDER, LZ ALBANY	NA		
777	CARTER	0535		SITE CO, LZ MACON GL075720	A5W38.51:30	SAME AS FOR LZ ALBANY	
777	DARGSECK	0535		GTA RATELO,LZ MACON	A5W38R51:30		
777	BOWER	0535		PATHFINDER, LZ MACON	NA		
NA	CARMEL	NA	NA	SITE CO, STANDBY RESERVE VIC DEKKAR STRIP	J4038. 37:30	SAME AS FOR LZ ALBANY	REMAIN WITH C 1 118TH (RESERVE) AT DEKKAR STRIP COMMITTED ON ORDER CALL SIGN WILL CORRESPOND WITH NAME OF LANDING SITE IF RESERVE IS COMMITTED
NA	KENFIELD	NA	NA	GTA RATELO, STANDBY RESERVE	J4038R..37:30		
NA	KEMP	NA	NA	PATHFINDER, STANDBY RESERVE	NA		
NA	CASSIE	NA		NONDUTY			

SUPPORTED UNIT: 1.188TH INF
LIFT UNIT: A/21ST AVN BN
ACL: 8 PAX
LZ TIME: 31 0600 MAR
DACOS: 53 AIR, 1:188TH INF, DFKK STRIP

TYPE ACFT: UH.1H
ARTY PREP AT LZ MACON AND ALBANY COMMENCES
H.20 MIN: LIFTED H.5 MIN.
ARTY PREP FIRED FROM GL 035725
CCP: GL 956696
RP: GL 015692

SIGNAL

1,188TH INF, RIGHT HALF. 47.00
A/21ST AVN BN: DEADLY SERPENT. 39.00
PFDR: DEKKAR CONTROL. 34.50
ALBANY CONTROL. 40.20
MACON CONTROL. 51.30
RESERVE. 37.30

(REFER TO GLOSSARY FOR ACRONYMS)

CLASSIFICATION

Figure A-1. Example format for operation plan.

PATHFINDER UNIT	21ST AVN BN PATHFINDER SECTION
SUPPORTED UNIT	1. 188TH INF
PERIOD	210001-212400 MARCH
OPERATION (AFLD LZ, DZ)	MACHER
....	(NAME OR LOCATION OF SITE)
RECORDER	PFC ROY S. COBB

FLT OR ACFT NO	TYPE ACFT	TIME COMM ESTAB	TIME		TYPE LOAD		REMARKS
			ARR	DPRT	DELIVERED	EVAC	
(EXAMPLE FOR AFLD OPERATION)							
S7M50	U-1A	0600	0610	0625	RATIONS	2W1A	
L8M82	U-1A	0605	0615		AMMO		DAMAGED PROPELLER TO DISPERSAL AREA
C6E91	C-123	0610	0620	0635	AMMO	4W1A	
(EXAMPLE FOR DZ OPERATION)							
F1Y11	C-7A	1705	1710		3 BUNDLES AMMO		
T3M34	U-6A	1715	NA				ABORTED 2 MILES OUT
B6M78	U-6A	1720	1725		3 BUNDLES RATIONS		
(EXAMPLE FOR HELICOPTER DZ OPERATION)							
A2Y41	UH-1D	1230	1235	1236	PERSONNEL		CONTACTED CONTROL CENTER WITH A FLIGHT OF FOUR
D3S32	UH-1D	1230	1235	1236	PERSONNEL		
E4C23	UH-1D	1230	1235	1236	PERSONNEL	4W1A	
H1Y14	UH-1D	1230	1235	1236	MORTAR AMMO		
J1P67	CH-47	1245	1250	1252	AMMO		

Figure A-2. Example formats for the LZ/DZ control record.

b. **Column Headings**.

(1) ***FLT OR ACFT NO***. This means "flight or aircraft number," so enter the flight or aircraft's radio call sign.

(2) ***TYPE ACFT***. This means "type aircraft," so enter the Army or Air Force aircraft model designation.

(3) ***TIME COMM ESTAB***. This means "time communication established," so enter what time the aircraft acknowledges contact (radio, visual, or both, as applicable).

(4) ***TIME (ARR, DPRT)***. Enter the time the aircraft arrives or when the first of the flight lands. Enter the aircraft's departure time, or the time the last of the flight clears the ground.

(5) ***TYPE LOAD (DELIVERED, EVAC)***. Enter what type of load the aircraft delivered (supplies, equipment, or personnel) or evacuated (supplies, equipment, or personnel).

A-3. TROOP-LEADING PROCEDURES

From the moment he receives an order to conduct a pathfinder operation, the pathfinder leader does his best to use the following troop-leading procedures:

a. **Section Warning Order**. Issue a section warning order, including--

(1) Roll call.

(2) A brief statement of the enemy and friendly situations.

(3) The mission.

(4) Chain of command and section structure.

(5) Individual uniform and equipment (if not in SOP).

(6) Equipment required.

(7) Time schedule to complete work priorities (state who must show up, and where and when they must show up).

(8) Specific instructions and attached personnel.

(9) Time hack.

b. **Tentative Plan**. Make tentative plan of operation.

(1) Study the map.

(2) Check the weather.

(3) Study the unit SOP.

(4) Make a quick estimate of the situation. Will you need extra personnel from the supported unit, other equipment or materiel, or additional communications resources?

(5) Begin planning.

c. **Movement and Coordination**. Arrange for movement and coordination.

(1) Arrange to move unit and inform second in command.

(2) Coordinate with ground and aviation units. Cover the ground tactical plan, the landing and unloading plan, the air movement plan, and the loading plan. Arrange for any extra people or equipment needed from the supported unit.

d. **Tentative Operation Plan**. Prepare a tentative operation plan.

(1) Reconnoiter when time permits (map, ground, air).

(2) Continue the estimate and receive recommendations.

(3) Complete the plan (work out details, formulate orders). Present the tentative plan to the supported unit commander or his staff. Prepare the final plan based on the desires of supported unit commander and on his final order.

(4) Issue section order (normally an oral order).

(5) Join supported unit.

(6) Rehearse and inspect (if time and terrain permit).

APPENDIX B

ARMY HELICOPTER SPECIFICATIONS

For pathfinders to help select landing sites, they must know the dimensions of Army aircraft that US Army active and reserve units could employ. The art in this appendix shows helicopters currently in use as of this manual's publication date.

B-1. OBSERVATION HELICOPTERS

This category includes the OH-58D Kiowa and the OH-6A Cayuse.

a. **OH-58D Kiowa**. Table B-1 shows specifications for the Kiowa; Figure B-1 shows the aircraft from three angles:

Rotor Diameter	35 FT	
Length:		
Rotor Operating	42 FT	2 IN
Blades Removed	33 FT	10 IN
Height to Top of Turret	12 FT	9 - 1/2 IN
Tread of Skids	6 FT	2 IN
Main Rotor:		
Disk Area	0.962 SQ FT	
Blade Area	38.26 SQ FT	
Clear Area Needed for Rotor	12.5 M	
TDP #1	25 M Diameter	

Table B-1. Specifications for the OH-58D (Kiowa).

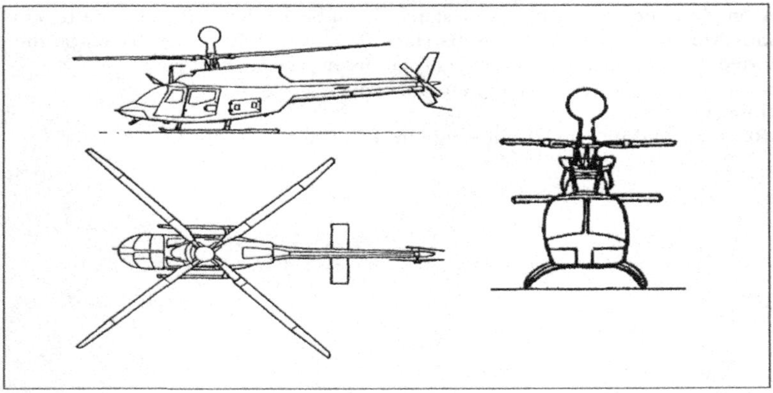

Figure B-1. OH-58D (Kiowa).

b. **OH-6A Cayuse**. Table B-2 shows specifications for the Cayuse; Figure B-2 shows the aircraft from three angles:

Rotor Diameter	26 FT 4 IN
Length:	
Rotor Operating	39 FT 3 - 3/4 IN
Rotor Folded	22 FT 9 - 1/2 IN
Height	8 FT 9 IN
Tread	6 FT 3 IN
Clear Area Needed for Rotors	9.3 M
TDP #1	25 M Diameter

Table B-2. Specifications for the OH-6A Cayuse.

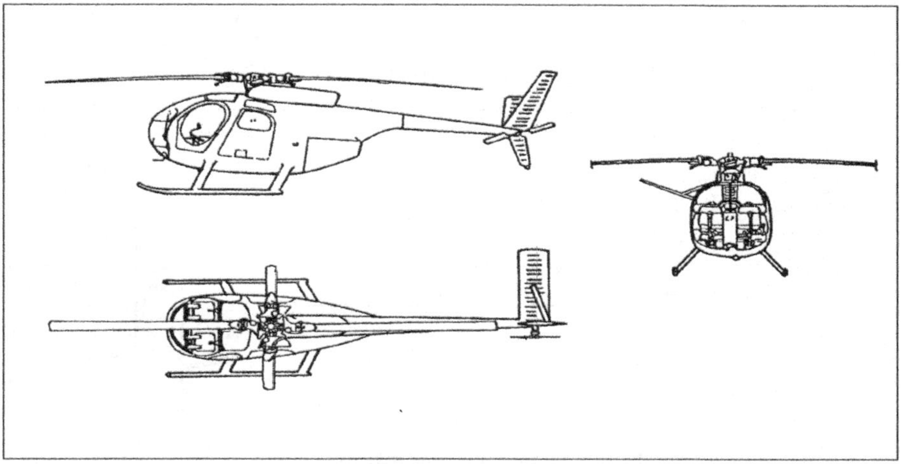

Figure B-2. OH-6A (Cayuse).

B-2. ATTACK HELICOPTERS

This category of helicopters includes the AH 1S Cobra and the AH 64A Apache.

a. **AH 1S Cobra**. Table B-3 shows specifications for the Cobra; Figure B-3 shows the aircraft from three angles:

Rotor Diameter	44 FT
Length:	
Rotor Operating	53 FT 1 IN
Fuselage	44 FT 9 IN
Span, Maximum Lateral	11 FT 8 IN
Height	11 FT 7 IN
Tread (Skids)	7 FT
Rotor Ground Clearance (Static)	7 FT 10 IN
Clear area needed for Rotors	16.18 M
TDP # 2	35 M Diameter

Table B-3. Specifications for the AH 1S (Cobra).

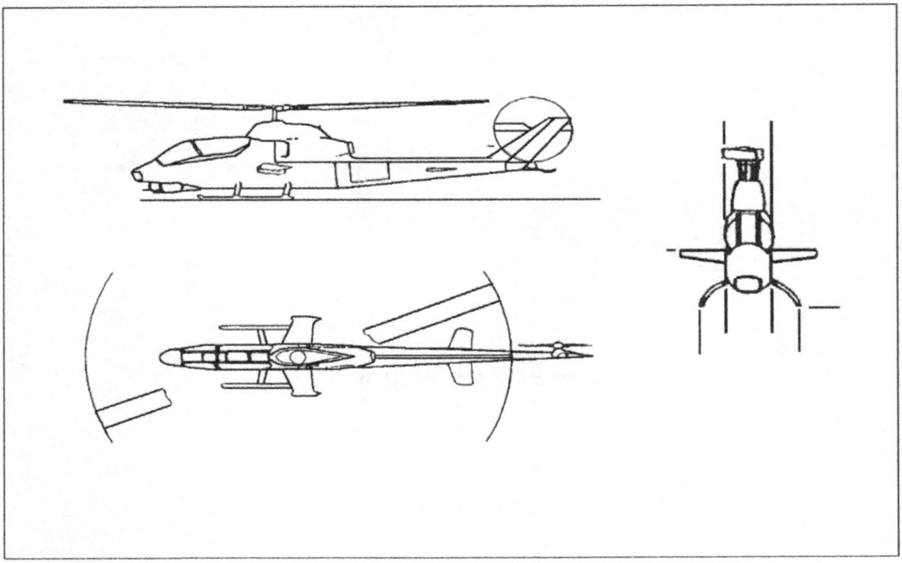

Figure B-3. AH 1S (Cobra).

b. **AH 64A (Apache)**. Table B-4 shows specifications for the AH 64A Apache; Figure B-4 shows the aircraft from three angles:

Rotor Diameter	48 FT	
Length:		
Rotors Operating	58 FT	3 - 1/8 IN
Rotors Static	57 FT	4 IN
Fuselage	48 FT	
Height	15 FT	3 - 1/2 IN
Clear Area Needed for Rotors		17.9 M
Minimum TDP without commander's approval is #3		50 M

Table B-4. Specifications for the AH 64A (Apache).

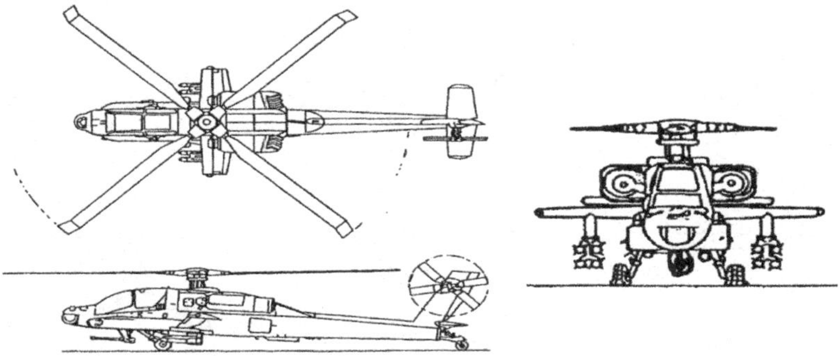

Figure B-4. AH 64A (Apache).

B-3. UTILITY HELICOPTERS

This category of helicopters includes the UH-1H Iroquois and the UH-60A Blackhawk.

a. **UH-1H Iroquois**. Table B-5 shows specifications for the Iroquois; Figure B-5 shows the aircraft from three angles:

Rotor diameter	48 FT
Length:	
Rotors Operating	57 FT 1 IN
Rotors Static	57 FT 1 IN
Fuselage	41 FT 10 - 3/4 IN
Span, Maximum Lateral	9 FT 4 IN
Height	14 FT 6 IN
Tread	8 FT 6 - 1/2 IN
Ground Clearance, Static Against Stops	6 FT 6 IN
Clear Area Needed for Rotors	17.4 M
TDP # 2	35 M Diameter
Allowable Cargo Load	4,000 LB

Table B-5. Specifications for the UH-1H (Iroquois).

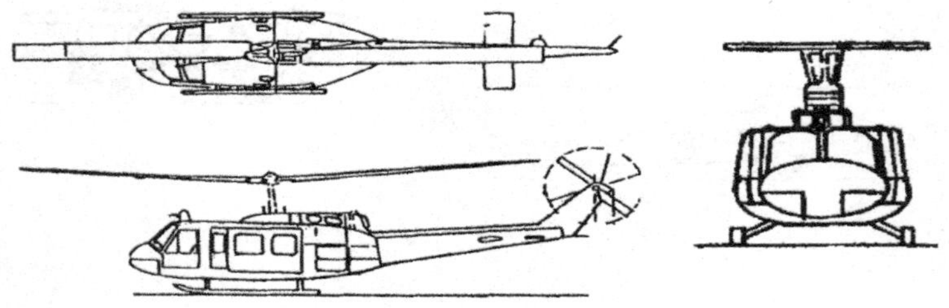

Figure B-5. UH-1H (Iroquois).

b. **UH-60A Blackhawk**. Table B-6 shows specifications for the UH-60A Blackhawk; Figure B-6 shows the aircraft from above and from the left side:

Rotor Diameter	53 FT	8 IN
Length:		
Rotor Operating	64 FT	10 IN
Rotor Folded	64 FT	10 IN
Fuselage	50 FT	7 - 1/2 IN
Span, Maximum Lateral	9 FT	8 - 1/2 IN
Height	16 FT	5 IN
Tread	8 FT	10 - 1/2 IN
Ground Clearance, Static, Against Stops	8 FT	9 IN
Clear Area Needed for Rotors	19.5 M	
TDP # 3	50 M Diameter	
Allowable cargo load	8,000 LB	

Table B-6. Specifications for the UH-60A (Blackhawk).

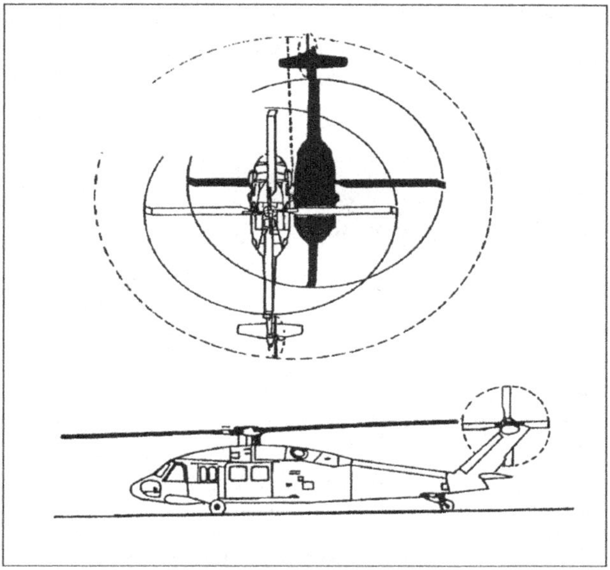

Figure B-6. UH-60A (Blackhawk).

B-4. CARGO HELICOPTERS

This category of helicopters includes the CH-47B/C and the CH-47 B/D Chinooks. With slingload, cargo helicopter TDP is #5 (100-meter diameter); without slingload, it is #4 (80-meter diameter).

a. **CH-47B/C Chinook**. Table B-7 shows specifications for the CH-47B/C Chinook; Figure B-7 shows the aircraft from three angles:

Rotor Diameter	60 FT	
Length:		
Rotors Operating	98 FT	10 - 3/4 IN
Rotors Folded	50 FT	9 IN
Height (Overall)	18 FT	11 - 1/2 IN
Tread	11 FT	11 IN
Rotor Ground Clearance:		
Static Forward	7 FT	4 - 3/4 IN
Idling Forward	10 FT	11 IN
Clear Area Needed For Rotors	30.4 M	
TDP # 4	80 M Diameter	
Allowable Cargo Load	21,000 LB	

Table B-7. Specifications for the CH-47 B/C (Chinook).

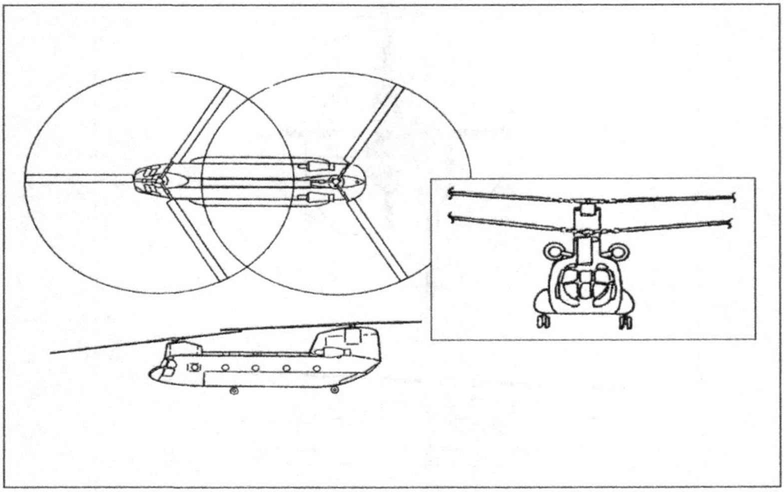

Figure B-7. CH-47B/C (Chinook).

b. **CH-47 B/D Chinook**. Table B-8 shows specifications for the CH-47B/D Chinook; Figure B-8 shows it from three angles:

Rotor Diameter	60 FT	
Length:		
Rotors Operating	98 FT	10 - 3/4 IN
Rotors Folded	51 FT	
Fuselage	51 FT	
Height (Overall)	18 FT	7 - 13/16 IN
Tread	11 FT	11 IN
Rotor Ground Clearance:		
Static Forward	7 FT	10 - 5/8 IN
Idling Forward	11 FT	7/8 IN
Clear Area Needed For Rotors	30.4 M	
TDP # 4	80 M Diameter	
Allowable Cargo Load:		
Forward Hook	17,000 LB	
Center Hook	26,000 LB	
Aft Hook	17,000 LB	
Any Two Hooks	25,000 LB	

Table B-8. Specifications for the CH-47 B/D (Chinook).

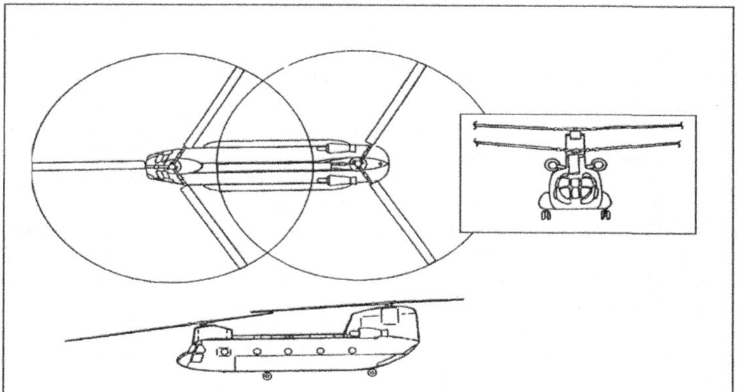

Figure B-8. CH-47 B/D (Chinook).

APPENDIX C

DIGITIZATION SUPPLEMENT

This appendix does not change existing doctrine, tactics, techniques, or procedures for employing pathfinder teams and elements. It just introduces a vision of how digitization will affect pathfinder teams.

C-1. DEFINITION

"Digitization" refers to the ongoing effort to integrate the Army's diversified battlefield operating systems into a network where leaders and soldiers can share and interchange current information about the battlefield environment for immediate or later use. FM 101-5-1 defines digitization as "*The near real-time transfer of battlefield information between diverse fighting elements to permit a shared awareness of the tactical situation.*" Accelerated operational tempo, instant communications, and immediate response times will characterize the digital environment.

a. Digitization is based on advancements in computers, communications, satellite navigation, and OTN vision and sighting capabilities. Satellite navigation, digital communications, and digital mapping will allow leaders and soldiers to share a common, digital picture of the battlespace environment in real time.

b. New computer capabilities will allow leaders to decisively concentrate their battlefield combat power. High-speed digital data exchange, the simultaneous display of intelligence data to leaders at all levels, and the rapid exchange of targeting data from sensors to shooters power this increased effectiveness.

c. New OTN capabilities enhance the ability to detect, identify, and engage targets during limited visibility. They also increase accuracy and control of fires, which reduces the risk of fratricide.

C-2. PURPOSE, ADVANTAGES, AND CAPABILITIES

The main purpose of digitization is to provide access to common, current data about friendly and enemy forces. Leaders and soldiers collect these data through networks of sensors, CPs, processors, and weapons platforms. Every leader and soldier will have immediate access to relevant information. All soldiers will know what is happening around and among them. Other purposes of digitization include the following:

a. **Information**. To provide an integrated digital information network that will improve information flow, increase command and control, support battlefield fire and maneuver, and give leaders (decision makers), soldiers (shooters), and supporters customized information they need in order--

(1) To make the decisions necessary to prevail in any campaign.

(2) To share vital battlefield information (friendly and enemy) rapidly, if not instantly, rather than by slow voice radio or even slower liaison efforts.

(3) To act on information faster than the enemy does.

(4) To win the information war against any adversary.

(5) To use common data collected through the digitized network

b. **Situational Awareness**. To achieve near real-time situational awareness at all times and at all levels. This will allow leaders and soldiers to collect, share, and use relevant information at once.

c. **Operational Awareness**. Greater situational awareness leads to increased *operational awareness*.

d. **Operational Tempo**. A greater grasp of the big picture (operational awareness) enables leaders and soldiers to set and maintain an increased *operational tempo,* which will help in overwhelming the enemy.

e. **Lethality, Accuracy, and Survivability**. To enhance the effectiveness and combat power of a force comprised of fewer and smaller units, so it can execute its missions with greater lethality, accuracy, and survivability in an environment characterized by an accelerated operational tempo, instant communications, and immediate response times.

f. **Planning and Preparation**. To streamline and support mission planning, preparation, and execution. To do so by providing critical information to every leader at the same time. Having earlier access to needed information compresses the mission planning and preparation cycles.

g. **Effectiveness**. To increase the pathfinder team's effectiveness in conducting pathfinder operations for its parent unit.

h. **Firepower**. To synchronize direct and indirect fires more effectively. To mass the effects of dispersed firepower. This replaces the existing way of fighting that relies on physically massing weapons and forces.

i. **Decentralization**. To support more decentralized operations, thereby improving the effectiveness of team- and element-level operations.

j. **Battlespace**. To maintain a clear, accurate, simultaneous, and appropriate vision of the battlespace at each echelon--from soldier to commander, and to enhance the mission capabilities needed in a multidimensional battlespace.

k. **Fratricide**. To minimize fratricide.

l. **Signature**. To reduce the team's signature on the battlefield.

m. **Navigation**. To improve navigation.

n. **Linkup**. To improve linkup.

GLOSSARY

A/C	aircraft
acc load	Cargo suspended below a slingload
ACL	allowable cargo load
ACP	air control point
ADS	aerial delivery sling
AF	Air Force
AFB	Air Force base
aft cargo hook	A cargo hook located aft of the center or main cargo hook
aft	To the rear, behind, or toward the tail of the aircraft
AGL	above ground level
AH	attack helicopter
ALO	air liaison officer
alt	altitude
AM	amplitude modulation
AMC	Air Mobility Command
anemometer	an instrument for measuring and indicating the force or speed of the wind
AO	area of operations
apex fitting pin	A headed pin that extends through the two ends of the apex fitting-clevis, retained in place by a bolt and nut apex fitting spacer (an hourglass-shaped bushing that fits over the apex fitting pin), causing the apex fitting to stay centered on the cargo hook
apex fitting	The uppermost point (clevis, shackle, ring, or loop) that gathers the sling tension members and attaches to the cargo hook
apex ring	(See apex fitting.)
apex shackle ftg	(See apex fitting.)

apex shackle	(See apex fitting.)
appr azimuth	The direction of an aircraft's approach to the landing point, usually expressed in degrees and measured clockwise from a reference point
arr	arrival
ATA	actual time of arrival
ATC	air traffic controller
ATD	actual time of departure
AWADS	all-weather aerial delivery system
AZAR	assault zone availability report
basket hitch	A type of attachment in which a flat web nylon strap is formed in the shape of the letter "U" with the two ends connected side-by-side
basket leg	The cable on a pallet sling that encircles the load
becket	A short length of rope with an eye on one or both ends, used to secure items such as eyelets or coils of rope
bight	A formed loop in a rope or cable
blivet	A rubber storage bag, transportable by air or ground, that holds 500 gallons of fuel or water
breakaway technique	Use of material with a low breaking strength, such as tape or cotton webbing, to temporarily restrain the sling equipment from becoming entangled on the load as the helicopter puts tension on the sling
breaking strength	Resistance to breaking, commonly measured in pounds. (Refers to tensile strength.)
brownout	A loss of orientation with respect to the horizon, caused by blowing sand, dust, or dirt
CA	combat assault
cable	A suspended external cargo sling system. Cargo hook attached to the bottom of the fuselage by a system of cables
CAPES	chemiluminescent light-assisted personnel exit system
cargo hoist	A device used to raise, lower, or winch cargo

cargo hook keeper A spring-loaded device used to prevent the apex fitting from sliding off the cargo hook load beam

cargo hook loading pole Also called a "shepherd's hook," with a hook on one end, which an aircrew member uses to pick up the apex fitting and connect it to the cargo hook. Used when the hookup team cannot stand on top of the load

cargo hook A suspension hook device on which the apex fitting is placed to transport an external load

cargo net A meshed arrangement of straps or cords that have been knotted or woven together at regular intervals and used to transport supplies and equipment

cargo sling Consists of multiple tension members which, when properly rigged, enable aircrew to suspend cargo and equipment under a helicopter

CARP computed air release point

castle nut A hexagon-shaped nut with a slot in each of its six sides and secured in place by a cotter pin

CC control center

CCP communications checkpoint

CDS container delivery system

cellulose A cushioning or padding material used to protect the load or sling legs

center (main) cargo hook A cargo hook located in the center of the helicopter

center of balance Location where an item would balance if it were possible to support the load at that point

center of gravity (See center of balance.)

centerline An imaginary line that extends lengthwise and bisects the bottom of the helicopter fuselage

CG center of gravity

CH cargo helicopter

chemiluminescent light A plastic tube filled with a liquid chemical and a glass vial that when activated produces a glowing light known as chemiluminescence

choker hitch A knot formed by placing a strap or cord around an item and passing the free end of the strap or cord through the loop formed by the other end clevis A "U"-shaped yoke device with a bolt or pin through the two ends

clinometer Any of various instruments used to measure elevation or incline

connector link A metal link device that joins the ends of a nylon web loop together, such as used on the web ring of the 15,000-pound capacity multileg sling

constant power The point at which a helicopter engine produces a fixed level of performance

CONUS continental United States

CP command post

CRC control and reporting center

dep departure

det detachment

DF direction finding

down wash Also called "rotor wash." High-velocity air movement under a hovering helicopter

downslope Downhill; descending; a slope that lies downward

downwind In the direction that the wind is blowing

drawbar A beam across the rear of a vehicle or aircraft to which implements are hitched

drop-off A very steep or perpendicular descent

DS direct support

DTG date-time group

dual-point load Cargo or equipment rigged and suspended from two cargo hooks

DZ drop zone

DZSO drop-zone safety officer

DZST drop-zone support team

DZSTL drop-zone support team leader

ECCM electronic countercountermeasures

ECM electronic countermeasures

end bar The removable side of a connector link

energy-dissipating material Artificial cardboard packing material. In parachute operations, protects equipment by dissipating shock or energy when the package lands

ETA estimated time of arrival

EW electronic warfare

exit path The track or course of an aircraft departing the landing point

ext extracted

external air transport Movement of supplies or equipment that is rigged and suspended from a helicopter cargo hook

external load Consists of supplies or equipment properly rigged with either one or more slings, cargo bags, or cargo nets

eye sock A loop formed at both ends of a sling leg

EZ extraction zone

FB Fort Benning

FEBA forward edge of the battle area

field-expedient drogue device Any device locally fabricated to connect to an external load for the purpose of increasing the stability of the load

field-expedient pendants Individual slings or sling legs used as a vertical riser or pendant

field-expedient static-discharge wand A device fabricated from readily available material to discharge the helicopter's static electrical charge

FLA frontline ambulance

flat web nylon A flat nylon strap or sling fabricated in either concentric loops or single lengths

floating web keepers (See sliding keepers.)

FM frequency modulated

FOC flight operations center

forward cargo hook A cargo hook located forward of the center or main cargo hook

forward slope Ground whose surface forms an upward angle or incline with the horizon

FRAGO fragmentary order

free-swinging cargo hook A cargo hook attached to the end of a pendant or cable instead of attached directly to the helicopter fuselage

fuselage The central body portion of a helicopter, which accommodates the crew, passengers, and cargo

fwd forward

GAIL glide angle indicator light

GCA ground-controlled approach

glide angle indicator light (GAIL) A device that emits a three-color light beam that indicates a safe glide path over approach obstacles onto a landing site

glide path angle The helicopter's angle of approach to the landing point

GMRS ground-marking release system

grabhook A hook for grabbing (as the links of a chain)

ground crew Personnel on the ground who help prepare and rig loads, guide the helicopter, and connect the load to the helicopter

GSI glide slope indicator

GTA ground-to-air

GUC ground unit commander

GWT gross weight

HAHO high-altitude, high-opening

HALO high-altitude, low-opening

handheld Designed to be operated while held in the hand

HEAT helicopter external air transport

HLZ helicopter landing zone

HMMWV high-mobility, multipurpose wheeled vehicle

IAW in accordance with

immediate mission A task or support mission that arises during the course of a battle and which by its nature cannot be planned in advance

INR internal net recorder

ins inserted

JAAT joint air attack team

JP jet petroleum

keeper A device that holds something in position such as an apex fitting on the helicopter cargo hook

KIAS knots indicated airspeed

knotless nylon cord cargo net A cargo net manufactured from braided nylon cord joined by a knotless intersection method

landing point A point within a landing site where one helicopter can land

landing site A site within a landing zone that contains one or more landing points. This is also known as the touchdown point

landing zone Any specified area used for the landing of aircraft

lanyard A short piece of rope or cable used to fasten something or an extension of a shorter device

lashing straps A flat-web nylon strap with a D-ring on one end. Used to lash equipment together

latch bar The movable top part of the pallet sling tensioner

lateral slope Ground whose surface forms an angle oriented from either right to left or left to right with the horizon

lifting eye A loop in a cable, rope, or other device used to lift an object

lifting leg hook A hook on the end of a sling leg, usually used with a hoisting sling

lifting point/lifting provision An integral part of the equipment, commonly called a lug, eye, shackle, or ring that provides a means of attaching a sling for safe lifting

lifting shackle A ring or clevis that provides a means of attaching sling for safe lifting liftoff A vertical takeoff by an aircraft or a rocket vehicle or a missile

LIN line item number

load beam A structural member of the cargo hook that supports the weight of the external load

load binders A locking device used to tighten and secure

load oscillation The swinging movement of an external load

locking detent A slot in the pallet sling tensioner stirrup that engages the release knob assembly on the latch bar

low-response, external cargo sling system A cargo hook attached to a suspension frame below the fuselage by a system of cables and pulleys

lunette The towing eye on the trailer drawbar

LZ landing zone

MAC Military Airlift Command

main cargo hook winch A mechanical drum with a cable attached to raise and lower the cargo hook

manual release knob or lever A lever, knob, or other device that, when actuated by either the ground crew or aircrew member, causes the cargo hook to open

mean effective wind The average wind speed from the ground to drop altitude

MEDEVAC medical evacuation

METT-T mission, enemy, terrain, troops, and time available

MEW mean effective wind

MK 100 (Mark 100) The designation given to the pallet sling that can accommodate a pallet load 48 to 70 inches high

MK 86 (Mark 86) The designation given to the pallet sling that can accommodate a pallet load 29 to 40 inches high

MOGAS motor gasoline

multileg-pole pendant A reach pendant that has one or more removable sling legs

NATO North Atlantic Treaty Organization

NAVAID navigation aid

NBC nuclear, biological, and chemical

NCO noncommissioned officer

night vision goggles An image-intensification device that improves visibility in low light situations by amplifying available light

NLT not later than

no-load condition A situation in which the sling legs are under no tension

no-sew fitting A web keeper secured in place by a locking fork

NVD night vision device

nylon donut A ring formed by joining two ends of a flat, web-nylon strap using a Type IV link or two-point link connector

OPORD operation order

oscillate To swing back and forth like a pendulum

overfly To fly over; to pass over in an airplane

pendant cargo hook A cargo hook that is connected to the end of a strap and suspended below the fuselage

pendant A releasable swivel hook. That is, a high-strength sling constructed of multiple plies of nylon webbing with a remotely operated, manually released, full 360-degree swivel hook

permanent keeper A web keeper that encircles all of the loops of a nylon web strap and is either sewn or otherwise retained in place

phraseology A manner of organizing words and phrases into longer elements; a choice of words

PI point of impact

PIBAL pilot balloon

pickup zone Landing site designated where an aircraft picks up passengers or cargo

piggyback load Cargo suspended below a slingload such as a cargo net carried beneath a howitzer

POL petroleum, oils, and lubricants

port side — The left-hand side (as one looks forward)

preclude — To make impossible by necessary consequence; to rule out in advance

preplanned mission — A task or support in accordance with a program planned in advance of the operation

proword — Procedure word; a word or phrase limited to radio telephone procedure used to facilitate communication by conveying information in a condensed standard form.

PSP — perforated steel planking

PZ — pickup zone

QSTAG — quadripartite standardization agreement

quick-fit strap fasteners — Metal, V-shaped device with a double bar and a friction-grip crossbar. Used with lacing straps

racetrack — To fly in an oval flight pattern over a drop zone or point of impact

RAM — raised-angle marker

reach pendant — Braided fabric rope assembly with an attached, stiffened reach tube and a loop on each end

reach tube — A length of tubing that encases the braided rope portion of a reach pendant

receiving unit — A unit receiving the transported supplies and equipment

relative humidity — The ratio of absolute humidity to the maximum possible density of water vapor in the air at the same temperature

rendezvous point — A clearly defined and visible point or assembly area for the ground crew during helicopter operations

retainer strap — Nylon web loop interwoven on the web ring of the 15,000-pound capacity multileg sling that retains the sling legs on the lower part of the web ring

retention pins — Metal devices used to secure objects such as clevises or shackles to the vehicle frame

rope lay — An individual grouping of wire strands woven or twisted together, which when braided together form a cable

rotor wash — (See down wash.)

RP — release point

safe working load The greatest capacity of a lifting device used in helicopter external load operations

SATB simulated airdrop training bundle

SIGINT signal intelligence

signalman A person who signals or who works with signals. A member of the ground crew who communicates with the pilot by means of arm-and-hand signals

SINCGARS single-channel, ground and airborne radio system

single-point load Cargo or equipment rigged and suspended from one cargo hook

SKE station-keeping equipment

sliding keeper A web keeper located near the ends of a flat web strap that encircles all of the loops and acts as a cinch

sling leg An individual tension member of the sling set

sling load Consists of supplies or equipment properly rigged with either one or more slings, cargo bags, or cargo nets

sling-to-clevis attachment Connecting a flat web nylon strap to a lift provision by means of inserting a clevis through the loop on the end of the strap and attaching the clevis to the provision

SOI signal operation instructions

SOP standing operating procedure

spreader bar A device that prevents the sling legs from bearing against the load, which could cause damage

STANAG standardization agreement

starboard side The right-hand side (as one looks forward)

static discharge wand crewman A member of the ground crew responsible for discharging the static electricity from the helicopter. Also called "static probe crewman."

static electricity A electrical charge produced and stored in the helicopter fuselage during flight

STT special tactics team

supported unit A unit requesting the mission to transport supplies and equipment

swage stops Short lengths of metal tubing crimped on the pallet sling cables

tandem load An external load consisting of two pieces of equipment joined together, such as a prime mover and a howitzer

TDP touchdown point

tensioner stirrup The fixed body of the pallet sling tensioner

tensioner A locking device on the pallet sling cables used to adjust the overall length of the cable

theodolite A surveyor's instrument for measuring horizontal and, usually, also vertical angles

tie-down provision An integral part of the equipment commonly called a lug, eye, shackle, or ring that provides a means of attaching a hook, strap, or cable to restrain equipment during shipment

tie-down strap A device consisting of a flat web nylon strap with hooks on one or both ends and a ratchet device used to secure items

TOT time on target

touchdown hover A helicopter landing with only part of the landing gear positioned on the ground

TOW tube-launched, optically tracked, wire-guided missile

transit To pass across

TSC training support center

TTB tactical training bundle

turbo meter An instrument for measuring ground wind speed

turnaway the act of refusing admittance

turnout An act of turning out

two-point link connector A device used to connect two ends of a flat web nylon strap together

Type III nylon cord A thin rope made of several strands of Type III nylon woven together

Type IV link connector A device used to connect two ends of a flat web nylon strap together

Type X sling A nylon aerial delivery sling consisting of two, three, or four consecutive loops of Type X nylon webbing

Type XXVI sling A nylon multiloop-line sling consisting of two or more consecutive loops of Type XXVI nylon webbing

UH utility helicopter

UHF ultra-high frequency

upslope A slope that lies upward; uphill

upwind The direction from which the wind is blowing

USAF United States Air Force

USMC United States Marine Corps

VAPI visual approach-path indicator

vertical pendant A high-strength sling or strap used to increase the clearance between the helicopter and the load

vertical riser A suspension member oriented in a vertical direction

VHF very high frequency

VIRS verbally initiated release system

VMC visual meteorological conditions

V-shaped approach path The area within a designated arc or sector measured outward from the center of the landing point

web ring connector bar A metal connecting link that joins the two ends of the web ring strap on the 15,000-pound capacity, multileg sling

web ring strap A nylon web loop that forms the web ring on the 15,000-pound capacity, multileg sling

web ring The apex fitting of the 15,000-pound capacity, multileg sling set

whiteout A loss of orientation with respect to the horizon caused by blowing snow, the sun reflecting on the snow, or an overcast sky

windchill factor The cooling effect of moving air

REFERENCES

SOURCES USED

These are the sources quoted or paraphrased in this publication.

Army Publications

FM 10-450-3	Multiservice Helicopter Sling Load: Basic Operations and Equipment. 10 April 1997.
FM 21-60	Visual Signals. 30 September 1987.
FM 24-18	Tactical Single-Channel Radio Communications Techniques. 30 September 1987.
FM 24-24	Signal Data References: Signal Equipment. 29 December 1994.
FM 57-220	Static Line Parachuting Techniques and Training. 19 August 1996.
FM 101-5-1	Operational Terms and Graphics. 30 September 1997.
TM 9-1370-206-10	Operator's Manual: Pyrotechnic Signals. 31 March 1991; with Change 1, 10 March 1992.

Air Force Instructions

AFI 11-231	Computed Air Release Point Procedures. 1 July 1998.
AFI 11-410	Personnel Parachute Operations. 1 March 1997.
AFI 13-217	Assault Zone Procedures. 1 June 1999.

DOCUMENTS NEEDED

These documents must be available to the intended users of this publication.

Army Publications

FM 3-97.6	Mountain Operations. 28 November 2000.
FM 90-3	Desert Operations. 24 August 1993.
FM 90-4	Air Assault Operations. 16 March 1987.
FM 90-5	Jungle Operations (How to Fight). 16 August 1982.

Army Forms

AF Form 3823	Drop Zone Survey. 1 February 1994.
AMC Form 168	Airdrop/Airland/Extraction Zone Control Log. 1 December 1992.
DA Form 7461-R	Internal Net Record.
MAC Form 339	Drop Zone Survey. February 1989.

*International Agreements**

QSTAG 585	Marshaling Helicopters in Multinational Land Operations. 23 April 1981.
STANAG 2863	Navigational and Communication Capabilities for Helicopters in Multinational Land Operations. 26 September 1988.
STANAG 3117	Aircraft Marshaling Signals. 17 October 1985.
STANAG 3281	Personnel Locator Beacons. 3 April 1978.
STANAG 3570	Drop Zones and Extraction Zones--Criteria and Markings. 26 March 1986.
STANAG 3619	Helipad Marking. 10 Jul 1980.

Internet Web Sites

Some of the documents listed in these References may be downloaded from Army websites:

U.S. Army Publishing Agency (USAPA)
http://www.usapa.army.mil

Army Doctrine and Training Digital Library (ADTDL)
http://www.adtdl.army.mil

Air Force Publishing
http://afpubs.hq.af.mil/

* To requisition copies of standardization agreements, send a completed DD Form 1425 to Naval Publications and Forms Center (NPFC), 5801 Tabor Avenue, Philadelphia, PA 19120.

INDEX

FM 3-21.38 (FM 57-38)
1 OCTOBER 2002

By Order of the Secretary of the Army:

ERIC K. SHINSEKI
General, United States Army
Chief of Staff

Official:

Joel B. Hudson

JOEL B. HUDSON
Administrative Assistant to the
Secretary of the Army
0224801

DISTRIBUTION:

Active Army, US Army Reserve, and Army National Guard: To be distributed in accordance with Initial Distribution Number 110401, requirements for FM 3-21.38.

Made in the USA
Monee, IL
04 November 2026

67384042R00095